Routledge Revivals

Resource and Environmental Effects of U.S. Agriculture

Originally published in 1982, this report explores long-term trends in demand for U.S. agricultural production, energy prices and agricultural technologies and their effect on natural resources such as land and water in the United States. Crosson and Brubaker also discuss possible policy modifications in order to lessen the environmental impacts expected to emerge from these trends. This title will be of interest to students of Environmental Studies.

W0114752

Resource and Environmental Effects of U.S. Agriculture

Pierre R. Crosson and Sterling Brubaker

First published in 1982
by Resources for the Future, Inc.

This edition first published in 2016 by Routledge
2 Park Square, Milton Park, Abingdon, Oxon, OX14 4RN
and by Routledge
711 Third Avenue, New York, NY 10017

Routledge is an imprint of the Taylor & Francis Group, an informa business

© 1982 Resources for the Future, Inc.

Publisher's Note
The publisher has gone to great lengths to ensure the quality of this reprint but points out that some imperfections in the original copies may be apparent.

Disclaimer
The publisher has made every effort to trace copyright holders and welcomes correspondence from those they have been unable to contact.

A Library of Congress record exists under LC control number: 82047984

ISBN 13: 978-1-138-96139-5 (hbk)
ISBN 13: 978-1-315-65980-0 (ebk)
ISBN 13: 978-1-138-96140-1 (pbk)

First published in 1982
by Resources for the Future, Inc.

This edition first published in 2016 by Routledge
2 Park Square, Milton Park, Abingdon, Oxon, OX14 4RN
and by Routledge
711 Third Avenue, New York, NY 10017

Routledge is an imprint of the Taylor & Francis Group, an informa business

© 1982 Resources for the Future, Inc.

Publisher's Note
The publisher has gone to great lengths to ensure the quality of this reprint but points
out that some imperfections in the original copies may be apparent.

Disclaimer
The publisher has made every effort to trace copyright holders and welcomes
correspondence from those they have been unable to contact.

A Library of Congress record exists under LC control number: 82047984

ISBN 13: 978-1-138-96139-5 (hbk)
ISBN 13: 978-1-315-65980-0 (ebk)
ISBN 13: 978-1-138-96140-1 (pbk)

Resource and Environmental Effects of U.S. Agriculture

Pierre R. Crosson and Sterling Brubaker

RFF PRESS
RESOURCES FOR THE FUTURE

Resource and Environmental Effects of U.S. Agriculture

Resource and Environmental Effects of U.S. Agriculture

Pierre R. Crosson
Sterling Brubaker

RESOURCES FOR THE FUTURE / WASHINGTON, D.C.

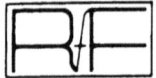

RESOURCES FOR THE FUTURE, INC.
1755 Massachusetts Avenue, N.W., Washington, D.C. 20036

Resources for the Future is a nonprofit organization for research and education in the development, conservation, and use of natural resources and the quality of the environment. It was established in 1952 with the cooperation of the Ford Foundation. Grants for research are accepted from government and private sources only on the condition that RFF shall be solely responsible for the conduct of the research and free to make its results available to the public. Most of the work of Resources for the Future is carried out by its resident staff; part is supported by grants to universities and other nonprofit organizations. Unless otherwise stated, interpretations and conclusions in RFF publications are those of the authors; the organization takes responsibility for the selection of significant subjects for study, the competence of the researchers, and their freedom of inquiry.

Research Papers are studies and conference reports published by Resources for the Future from the authors' typescripts. The accuracy of the material is the responsibility of the authors and the material is not given the usual editorial review by RFF. The Research Paper series is intended to provide inexpensive and prompt distribution of research that is likely to have a shorter shelf life or to reach a smaller audience than RFF books.

Library of Congress Cataloging in Publication Data

Crosson, Pierre R.
   Resource and environmental effects of U.S. agriculture.

   Includes index.
   1. Agriculture—Economic aspects—United States.
   2. Soil conservation—United States. 3. Agricultural ecology—United States.
   I. Brubaker, Sterling.   II. Title.
   HD1765 1982   333.76′0973   82-47984
   ISBN 0-8018-2920-8   AACR2

CONTENTS

vi

Tables

PREFACE

This report is the culmination of several years of research undertaken at Resources for the Future. The study originated out of concern
that prospective long-term trends in demand for U.S. agricultural production, in prices of energy and other key inputs, and in agricultural
technologies would put increasing pressure on the nation's land and
water resources and on the environment. It appeared to us that these
trends could shift the agricultural situation of the United States from
one of chronic surplus to one of recurring if not chronic scarcity.
Should such a shift occur, the nation likely will require major modification of the policies for resource and environmental management in agriculture which have evolved since the 1930s.

The study is addressed to analysis of the trends which over the
next several decades will determine the amount of pressure on the nation's
land and water resources, to evaluation of the resulting environmental
impacts, and dicussion of policy alternatives for dealing with them. With
the exception of interviews conducted as part of our research on tillage
and pest management technologies, irrigation and 208 planning, we have
not developed new sources of data. Rather we have relied on systematic
surveys of data and other information already available elsewhere. Nor
have we developed formal models of the various relationships studied. We
did, however, make use of Iowa State University's model of U.S. agriculture to project erosion and of a model developed at RFF for another purpose to transform projections of erosion to projections of sediment delivered. With these exceptions, the analysis is not cast in a formal
modeling mode. This would not have been appropriate, in our judgment,
given the broad focus of the research and its intent to identify and
assess new patterns of resource use and technology likely to emerge over
several decades.

The research was partially funded by grants from the Rockefeller
Foundation and the Environmental Protection Agency's Environmental

Laboratory at Athens, Georgia, as part of the Laboratory's program of research on Environmental Implications of Trends in Agriculture. We are grateful for their help.

Washington, D.C.   Pierre R. Crosson
June 1982   Sterling Brubaker

ACKNOWLEDGMENTS

Many persons read and made valuable comments on all or parts of this study. We particularly would like to thank George W. Bailey, Sandra Batie, Charles Benbrook, Daniel Bromley, Emery Castle, Lee A. Christensen, Fred T. Cooke, William H. Cross, Ray Frisbie, Donald R. Griffith, Maureen K. Hinckle, Lawrence W. Libby, John H. Perkins, David Pimentel, Kent Price, William Ramsay, Fred Sanderson, Frank W. Schaller, Robert Shulstad, Grant W. Thomas, John F. Timmons, G.B. Triplett, Jr., and R.D. Wauchope. Thanks go also to various people interviewed in the course of doing the research. They are cited in the text as appropriate.

We are grateful to all of these people for their help. Of course, they are blameless for the study's remaining defects.

Maybelle Frashure, ably assisted by John Mankin and Lorraine Van Dine, was responsible for typing the manuscript. We appreciate their uncomplaining (to us anyway) dedication to a sometimes tedious task.

P.R.C.
S.B.

Chapter 1

INTRODUCTION

Approaches and Caveats

In assessing future pressures on the nation's land and water resources and the environment, we have relied heavily on analysis and
projections of trends in three key variables: (1) demand, especially
export demand, for food and fiber; (2) prices or opportunity costs of
energy, fertilizer and water for irrigation; (3) agricultural technology.
The analyses and projections of the three variables are dealt with in
chapters 2, 3, and 4, and related appendixes. It seems useful at the
outset to state our approach to these projections and their meaning for
the study as a whole.

We view the trends in the three variables as the net outcomes of
numerous, often conflicting, forces which can be grouped in two cate-
gories--market and policy. Market forces and policy interact in that
market trends perceived to be socially undesirable may trigger policy
responses, and these may have unintended as well as intended effects on
market trends.

Our projections of the three variables present what we believe to
be a plausible yet somewhat conservative or pessimistic judgment of the
combined effect of future market forces and present policies on trends
in the variables. By "pessimistic" we mean that the projections are
more likely to overstate than understate future pressure on land and
water resources and resulting environmental problems. The projections
reflect present policies because we must first try to understand how
those policies will influence the future before discussing alternatives
to them.

If our judgment about future pressures on resources is wrong, we believe it most likely will be because we overestimate the rise in real prices or opportunity costs of energy, fertilizer, and irrigation water and underestimate the development of new technologies that relieve pressure on the land. There is a strong argument, cogently developed by Barnett and Morse (1963) and by Hayami and Ruttan (1971) and others, that emerging resource scarcities trigger both market and policy responses to find ways around the scarcities, thus permitting continued growth of output at constant or even declining real cost. This occurs because new management practices and technologies are developed which substitute more abundant resources for the more costly ones.

If real prices of energy, fertilizer, and irrigation water rise as we have projected, producers and government will have incentive to develop less costly substitutes for them. If yields increase at the relatively slow pace we have projected, the increased economic cost of agricultural land and mounting damages from erosion likely will stimulate increased research to develop new yield-increasing technologies.

Our projections of resource prices and yields were made in full awareness that market and policy responses to the problems that we foresee may lead to a different outcome. However, neither we nor anyone else has sufficient understanding of the institutional mechanisms which may trigger these responses to predict when or even if they will be forthcoming. Two considerations suggest, however, that certain key responses could be delayed a decade or more and be relatively weak when they do emerge. One is that modern agricultural technology, with its massive use of fertilizer and pesticides, was developed in large part to take advantage of cheap fossil fuel energy. Not only is the technology itself keyed to energy from fossil fuels, but the research establishment that developed the technology also is oriented to exploitation of this resource. As a consequence the scientific knowledge and technical skills needed to develop low-cost energy substitutes for fossil fuels--say through increasing the photosynthetic efficiency of main crops and enhancing the processes for biologically fixing nitrogen--are in short supply. And the supply of the necessary knowledge and skills cannot be quickly expanded.

Second, the market responses to increasing environmental damages may
be weak or nonexistent because these damages typically are not priced, so
knowledge of their quantitative importance is poor. Moreover, it is in
the nature of environmental costs that the people bearing them cannot
readily exact compensation from the people imposing them, and the lack
of clear price signals indicating the magnitude of environmental damages
also may confuse policy responses.

In structuring the study we aimed at comprehensiveness in two senses:
(1) we focused on the nation as a whole, although considerable attention
is given to regional and state issues; (2) we deal with all major environmental impacts of crop production, meaning erosion, fertilizer and pesticide pollution, and soil and water salinity associated with irrigation.
The research, being comprehensive and undertaken with a relatively modest
commitment of resources, necessarily is cast at a high level of general-
ity. The intent is to identify and assess the relative importance of
trends in crop production and technology that, from a national perspective, may have significant effects on the environment over the next several decades. Given this perspective and time frame, the conclusions of
the research necessarily are subject to greater uncertainty than would be
the case if the time span were shorter and the focus more narrow. The
conclusions, therefore, about trends in technology, resource use, environmental impacts, and policies are not intended to apply to specific local
situations. The conclusions should be meaningful and useful, however, in
providing insights as to emerging environmental problems of agriculture
and the effectiveness of alternative policies and institutional modes for
dealing with them.

Difficulties arising from uncertainty about the future are inherent
in studies such as this one. To narrow the uncertainty such studies frequently employ a set of projections for each important variable in an
attempt to bound the range of probable outcomes. We chose not to do this.
Had we done so the result would have been a proliferation of projections
and combinations of projections which would have served more to confuse,
and weary, than enlighten the reader. We believe the analysis underlying
each of our projections will enable readers to judge their reasonableness.
Those who think that our projections of future production and input prices

are too low and our projections of technological change too high will
conclude that we have underestimated future resource pressures and environmental problems. Those who believe we have erred in the opposite
directions will have a cheerier view of the future than we do.

Our analysis of trends points to rising real costs of agricultural
production and mounting environmental damage, especially that resulting
from erosion. We believe that the situation will call for fresh policies
to contain the costs and check the damages. However, we do not regard the
situation as posing a threat calling for immediate radical action. Moreover, for reasons given above, we believe we are more likely to have over-
rated the threat than to have underrated it. Accordingly, we see our-
selves, and expect to be seen by others, at a considerable distance from
the more grim visaged purveyors of gloom about the future of American
agriculture.

Although we conclude that both production costs and environmental
costs are likely to rise, we make no quantitative estimates of either.
This is unsatisfactory, both intellectually, and from the standpoint of
policy. Conclusions about the direction of costs are useful, but without
estimates of quantity also, judgments of importance inevitably are more
tentative than one would like. However, given the time frame and national
scope of our analysis and the fact that most environmental damages are
not priced, attempts to value those damages or to quantify production
costs would give a wholly spurious precision to the results. Those readers who believe quantification of production and environmental costs is
a sine qua non for useful discussion of the issues we address should stop
here.

Although we have not made specific projections of economic and environmental costs, we have projected the quantities of production of
main crops, the amounts of cropland and fertilizers farmers will demand,
the resulting amount of erosion from cropland, and sediment delivered to
lakes, rivers, reservoirs, and harbors. We also project the direction of
change in the amounts of pesticides used by farmers, but not the amounts
themselves. One's notion of the future magnitudes of these various quantities is crucial to one's judgment of whether pressures on the resource
base and environment will change much or little, and in which direction.

The projections do not translate into specific estimates of economic and environmental costs, but they provide an indication of whether emerging trends in agriculture portend future national problems.

Summary of the Report

The study begins with projections of production of wheat, feedgrains, soybeans and cotton. These crops occupy 70 to 75 percent of the nation's cropland and a significant proportion of its irrigated land. Their shares of fertilizers and pesticides used by farmers are comparable to their share of cropland. Moreover, these are the crops for which demand is increasing most rapidly.

Production of these crops for both domestic and foreign markets is projected from 1978/80 to 2010. The projections imply average annual production growth over this period of 1.8 percent for wheat, 1.6 percent for feedgrains, 2.6 percent for soybeans, and 1.0 percent for cotton. For comparison, annual production growth from 1969/71 to 1978/80 was 4.0 percent for wheat, 3.3 percent for feedgrains, 6.3 percent for soybeans, and 2.3 percent for cotton. In comparison with the 1970s, our projections are conservative.

Export demand is the dynamic element in projected production of these commodities, as it was in the 1970s. Growing world income and population, especially in developing countries, is expected to spur continued expansion of world trade in them, although at a slower rate than in the 1970s. Both China and the Soviet Union are expected to remain important net importers of grain, as is Japan of grain and soybeans. The European community likely will be a significant net exporter of wheat, but an importer of feedgrains and soybeans.

The U.S. increased its share of world trade in grains and soybeans in the 1970s, in part at least because of the devaluation of the dollar in 1971. We project constant U.S. shares at the levels reached in the late 1970s.

Our projections of domestic demand assume that for grains demand will grow in step with population, but that some increase in per capita consumption of soybeans will occur. The projection of cotton production, both for domestic consumption and for export, is based on a study by the U.S. Department Agriculture (USDA).

Our projection of domestic demand for feedgrains assumes that gasohol production will not generate a large derived demand for corn. Gasohol currently is competitive with gasoline only because of federal and state tax subsidies. Although the real price of gasoline likely will rise over the next several decades, coal, or perhaps wood, is likely to prove a more economical feedstock for liquid fuel production than grain.

We do not project specific prices for the commodities of interest. However, comparison of our projections with a set prepared by the USDA, which did include prices, suggests that our projections are consistent with real price increases of 25 to 30 percent over the next several decades.

The resource and environmental impacts of the projected amounts of production will depend in large part on the kinds of technologies farmers employ. We find it useful to distinguish among these technologies according to whether they are land-using or land-saving. Land-using technologies are marked by a relatively high ratio of land to non-land inputs, and land-saving technologies by the reverse. Other things the same, crop yields are higher with more land-saving technologies.

Farmers' choices among these technologies are conditioned by the relative prices of land and non-land inputs and by the productivities of each at the margin. From the end of World War II until the early 1970s prices of key land-saving inputs--fertilizers, energy and irrigation water--were relatively low and their productivity relatively high. These conditions, combined with government policies encouraging farmers to take land out of production, favored land-saving technologies, and crop yields increased at unprecedented rates.

After 1972, real prices of energy and fertilizer rose sharply, and there was evidence of increasing water scarcity in the west. In addition, the marginal productivity of fertilizer had fallen, and government policies to restrict land use became less binding. Land-saving technologies became less attractive as a result. Instead of diminishing as it had for more than two decades, land in crops increased about 60 million acres from 1972 to 1980. And the growth of yields slowed significantly.

The prospects for land-saving inputs suggest continuing increases in real prices and no major new sources of productivity growth. The implication is that farmers will continue to employ relatively more land-using technologies, as they did in the 1970s, and that crop yields will grow at the relatively slow pace of that decade.

This scenario regarding trends in crop production and technology provides the basis for projections of demand for cropland and fertilizers. By 2010 farmers would need an additional 60 to 70 million acres of cropland, bringing total crop acreage well above any previous amount in the nation's history. (Land in crops already was at a record high in 1981.) The amount of fertilizer applied would continue to increase, both in total and per acre, but at a much slower rate than in the 1950s and 1960s.

About 40 percent of the insecticides used on crops are used in cotton, with another 20 percent on corn. The amount used on cotton is likely to decline sharply, both because of a continuing shift of cotton acreage to Texas, where per acre use is low, and a decline in per acre use in the Mississippi Delta, where it currently is high. Climatic and economic conditions in Texas favor relatively low amounts of insecticides on cotton, and the state also has pioneered in developing integrated pest management (IPM) for cotton insects. IPM for cotton also is spreading in the Mississippi Delta, in the Southeast, and in California, the other major cotton producing state.

So far, IPM for corn has proved less attractive economically than for cotton, but it has potential for that crop also.

On balance, total insecticide use on cotton and corn is likely to decline sharply, and there is no evidence of offsetting increases on wheat, soybeans or other crops. The trend of insecticide use, therefore, is down.

The reverse is true for herbicides. Per acre use of these materials is likely to increase because of the spread of conservation tillage, a practice which for weed control relies more on herbicides and less on cultivation than conventional tillage.

The projections of resource use permit judgments about the severity of impacts on the environment. Except for a few "hot spots" around the country where nitrates in groundwater exceed Public Health Service stand-

ards, fertilizer pollution of water is not now a serious problem. The
increased use of fertilizers may exacerbate the problem in some areas, but
in a national perspective it is likely to remain of secondary importance.

The decline in quantity of insecticides used suggests that environmental damages from these materials will diminish. Herbicides generally
are not toxic to animals, and most are quickly degraded in the environment. Present evidence indicates they are not a major environmental
threat, either at present levels of use or at the higher levels projected
for the future. However, not all avenues by which herbicides may impact
the environment have been investigated. Moreover, there may be damage
threshholds above present levels of use which would be passed by the projected levels. In this event, the absence of compelling evidence of damage at present levels of use would be misleading in judging the future
threat. For these reasons, the increased use of herbicides should be
monitored carefully and research pushed to identify more precisely the
environmental impacts of these materials.

Soil and water salinity resulting from irrigation is a problem in
parts of the Colorado river basin, in the San Joaquin valley of California,
and in other scattered parts of the 17 western states. Irrigated acreage
is not expected to expand much over the next several decades, but the
severity of the salinity problem is as much, or more, a problem of water
management as it is of the amount of irrigated land. The salt content of
the soil tends to increase if it is not periodically leached out by rain
or irrigation water. This protects the leached soil but it increases the
salt content of irrigation return flows delivered downstream to other
users. Management of salinity thus is a perennial problem in arid irrigated areas. However, economically feasible techniques for dealing with
it are known. It is not expected to become a major problem over the next
several decades.

In the projected scenario erosion appears to present the principal
threat to the environment. Erosion damages are of two sorts: off-the-
farm, e.g. accelerated sedimentation of reservoirs from water erosion and
costs of clean-up from wind erosion; and on-the-farm in productivity losses. Farmers have little if any incentive to control off-farm damages of
erosion. While private bargaining between them and those damaged may

sometimes induce corrective action, this will not work unless the number of people is small and responsibility for the damage is easily established. When the numbers involved are large and the damage widely spread, private bargaining will not serve. These conditions make a prima facie case for public intervention to reduce off-farm damages.

The case for public intervention to reduce productivity losses of erosion is much less clear. The farmer has an interest in reducing these damages if the cost to him of doing so is less than the cost of the damages. The case for public intervention rests on the argument that protecting the social interest in the productivity of the land requires a greater conservation effort than the farmer will voluntarily undertake to protect his own interest. It is a market failure argument, and in our judgment it is not as easily made as is commonly believed. The strongest element in the argument is that society's responsibilities to future generations may require it to be more cautious than the market in using up an exhaustible resource--the land. The argument is complicated, however, by the fact that not only may technology substitute for the land, but often apparently "exhausted" land can be restored. Can society best meet its responsibilities to the future, assuming the market falls short, by additional soil conservation efforts, by developing and investing in soil restoration techniques, or by investing in technological substitutes for the soil? The list does not exhaust the possible policy alternatives, but it suggests the complexity of the policy issue.

Our projections of future erosion indicate a substantial increase by 2010, given the production and land use scenario for that year. Sediment delivered to the nation's waterways would about double. We conclude that erosion on this scale likely would be viewed as a major national problem, although we are unable to estimate the costs, either of off-farm damages or of losses of productivity.

Because erosion emerges as the principal environmental threat, the discussion of policies is focused mainly on erosion control. Traditional soil conservation policies of the USDA are reviewed as are the efforts of 6 states to plan for control of non-point pollution under section 208 of the Clean Water Act of 1972. The Act makes the Environmental Protection Agency (EPA) responsible for meeting water quality standards, but the

Agency has delegated 208 planning to the states. Those parts of the plans addressed to erosion control rely heavily on conservation practices long advocated by the USDA and on voluntary adoption of them by farmers.

Voluntarism traditionally characterized USDA soil conservation programs, and still does. The programs stress education, persuasion and financial incentives to induce farmers to adopt erosion control practices. Appraisals of the programs have found that much of the conservation effort has been applied to farms where erosion is not a serious problem. The USDA now proposes more precise targeting of conservation resources, and this should improve performance if acted upon. However, if erosion mounts as we have projected it, new approaches to control likely will be needed. Programs tieing participation in price support programs to adoption of soil conservation practices would be worth exploring, as would tax and regulatory approaches. And the EPA may have to take a sterner view toward state measures to control erosion if water quality objectives are to be achieved.

These various approaches seek to reduce erosion by inducing or requiring farmers to adopt erosion control practices. An alternative is to invest in research to develop higher yielding, less erosive technologies. If a faster rate of yield increase than we have projected could be achieved, pressure on the land would be diminished and the erosion threat eased. Research to extend the limits of conservation tillage, although not lessening the demand for land, could greatly reduce erosion per acre. Research to develop these sorts of technology, therefore, should be viewed as a policy alternative for dealing with the erosion problem. Historically, technological advance has been a prime instrument for extending the natural resource base. We should now actively explore its potential for relieving emerging pressures on our land and water resources.

References

Barnett, Harold, and Chandler Morse. 1963. Scarcity and Growth (Baltimore, Md., Johns Hopkins University Press for Resources for the Future).

Hayami, Yujiro, and Vernon Ruttan. 1971. Agricultural Development: An International Perspective (Baltimore, Md., Johns Hopkins University Press).

Chapter 2

PROJECTIONS OF PRODUCTION OF GRAINS,
SOYBEANS, AND COTTON IN THE UNITED STATES

Introduction

We are interested in the projections because the levels of production of these crops will determine in a major way the quantities of land, water, fertilizers, and pesticides that will be used in the United States. Alternative quantities of these resources, in turn, have varying economic and environmental consequences, the analysis of which is the main purpose of this study.

We deal with grains (wheat and feedgrains), soybeans and cotton for two reasons: (1) these crops consistently account for 70-75 percent of the land harvested in the United States and for high percentages of the fertilizers and pesticides applied; (2) if the growth of agricultural production in the United States over the next several decades puts substantial pressure on the demand for resources and on the environment, it will be because of the growth of production of these crops, particularly grains and soybeans. The growth in demand for all other agricultural commodities will respond primarily to growth in domestic population and income, and almost surely will be moderate. Under some conditions, however, export demand for grains, soybeans and cotton could increase enough to stimulate major increases in production of these crops.

Projections to 1990 and 2010

Table 2-1 shows the projections as well as actual figures for 1978-80. The U.S. Department of Agriculture (USDA) projections were made by

Table 2-1. U.S. Production, Export, and Domestic Use of Wheat, Feedgrains and Cotton, 1978/80 and Projections to 1990 and 2010
(millions metric tons)

Crop	1978	1979	1980	Average 1978/80	USDA 1990	RFF 2010
Wheat						
Prod.	48.3	58.1	64.5	57.0	77.1	98
Export	32.5	37.4	41.5	37.1	50.4	70
Dom. Use	22.8	21.3	22.9	22.3	26.4	28
Feedgrains[a]						
Prod.	221.1	238.8	198.7	219.9	282.0	354
Export	60.2	71.4	74.3	68.6	97.1	167
Dom. Use	157.2	161.9	155.8	158.3	184.9	187
Soybeans						
Prod.	50.9	61.7	49.4	54.0	72.1	120
Export[b]	27.7	32.9	29.5	30.0	33.8	76
Dom. Use	23.5	23.7	24.0	23.7	38.3	44
Cotton						
Prod.	2.4	3.2	2.4	2.7	2.7	3.5-3.9
Export	1.4	2.0	1.2	1.5	1.2	
Dom. Use	1.3	1.4	1.2	1.3	1.5	

Note: For 1978 the difference between production and the sum of exports and domestic use is the change in stocks. In the projections stock
changes are assumed to be zero.

Sources: 1978-1980 from U.S. Department of Agriculture, Foreign Agricultural Circular Grains, FG-4-81 (Washington, D.C., Jan. 28, 1981) for
wheat and feedgrains; USDA, Agricultural Outlook (Washington, D.C., March
1981) for soybeans and cotton. USDA projections for 1990 provided by
Leroy Quance, done in summer of 1980. They represent a "baseline" situation, the assumptions of which are given in the text. The projections
are preliminary and not official. More recent, and much higher, USDA
projections were published in the National Agricultural Lands Study, Final
Report (Washington, D.C., GPO, 1981).

[a]Corn and sorghum for grain, oats, and barley.

[b]The USDA projections are beans only. The 1978-80 figures and RFF
projections to 2010 are beans plus soybean meal and oil exports converted
to the bean equivalent.

the National Interregional Agricultural Projections (NIRAP) model, and
reflect a baseline situation. The principal assumptions underlying the
baseline projections are that: (1) U.S. population will grow at the same
rate from 1979 to 1985 as from 1970 to 1979 (.9 percent annually) and at
a slightly slower rate (.8 percent annually) from 1985 to 1990; (2) real
GNP will grow 3.0 percent per year from 1979 to 1990, slightly less than
in the 1970s; (3) export growth will be "moderate", i.e., less than in
the 1970s.

We accept the USDA projections as a reasonable set of outcomes for
1990. We think the projections of exports of feedgrains and cotton may be
somewhat low and those for domestic use of feedgrains a bit high, but in
neither case are they implausible.

There are no up-to-date USDA projections for 2010, so we made those
shown in Table 2-1 independently. The derivation of the projections is
described in the appendix to this chapter. Here we discuss some general
considerations relevant to evaluation of the projections.

Sources of Error in Projections

The projections game is one many can, and do, play, and there is no
firm basis for distinguishing the quality of the players. The range of
equally plausible outcomes is wide because the range of plausible assumptions determining outcomes is wide. Over periods of 20 or 30 years, small
differences in assumed rates of increase in population and per capita
income, in income elasticities of demand, or in rates of growth of food
production can yield large differences in projections of world trade, a
key element in our, and other, projections.

When projections depend upon extrapolation of trends--and all do to
some extent--problems arise if the trend is not clear. Which periods
should be selected to represent the trend if the data suggest it has
shifted over time? Ordinarily, the most recent period should be used,
but one frequently cannot be sure whether recent experience marks the
beginning of a new trend or only a momentary departure from established
trend. This is an especially difficult problem with trends in agricultural production because the data may reflect variations in the weather.
For example, a combination of unusually bad weather early in the period

and unusually good weather later can give a quite marked, and misleading, upward tilt to the trend in crop production.

Projections over periods as long as 20 or 30 years may also go wide of the mark because of the emergence of wholly unforeseen factors. Changes in dietary standards and habits could have a major impact on projections of food consumption and production. For example, our projections of feedgrain consumption in high-income countries assume that per capita consumption of meat in those countries will rise toward U.S. levels, thus increasing per capita consumption of feedgrains and soybean meal. Trends in meat, feedgrain and soybean meal consumption in those countries indicate this is a plausible assumption, but concern about the health effects of high-meat diets may dampen those trends, leading to slower than projected growth of demand for feedgrains and soybean meal.

The emergence of new uses for crops may have the opposite effect. The prospect for long-term increase in the real price of fossil fuels has stimulated growing interest in alternative sources of energy. The use of crops to produce ethanol for blending with gasoline to form gasohol is an example. At present, gasohol requires a subsidy to be competitive with gasoline, but continuing increases in the price of gasoline and progress in gasohol technology could eliminate this need. This is discussed below.

The development and spread of new, and unanticipated, production technologies also may cause outcomes to differ widely from projections. Unexpected breakthroughs in research on photosynthetic efficiency or in development of new seed varieties conceivably could lead to faster than projected growth in food production and slower than projected growth in import demand in importing countries. Given the present state of knowledge of these processes, such occurrences are not likely to affect five-to-ten-year projections, but they cannot be ruled out for longer periods. However, since there is no way of knowing when these developments may occur or what their production, consumption, and trade consequences may be, there is no point in trying to incorporate them in projections.

Policy changes also can cause projections to look foolish. Should the European Community (EC), for example, move toward more liberal trade and lower prices for agricultural commodities, the Community, a high cost producer, would increase its imports of grains substantially more than we

have projected. The U.S. certainly would capture a significant part of
this increased trade. We do not project that EC policies will change in
the indicated direction, although there is evidence of increasing concern
in the community about the high cost of present policies and the strength
of consumer interests seeking lower real prices of food is growing. Policy changes in other important trading areas or countries, of course, are
conceivable, although none that we have thought about are sufficiently
likely, in our judgment, to warrant inclusion in the projections. We may
be wrong, however. If so, then our projections will prove to be incorrect.

For all of these reasons we hold no brief that our projections are
necessarily more plausible than others that have been or might be done,
and we are confident that the projections will be wrong in ways we cannot
now foresee. We believe, however, that our projections are internally
consistent and reasonably in accord with present understanding about
trends and policies bearing on U.S. production of grains, soybeans and
cotton. That is all we claim for them.

Comparisons with Others

Table 2-2 provides a basis for comparing our projections with those
by the USDA (from Table 2-1) and by Martin Abel (1982). For comparison
with our projections, we extrapolated those of the USDA and Abel assuming
that the annual increments of production from 1977/79 to their target
dates would continue to 2010. Thus the projections for 2010 labeled USDA
and Abel in Table 2-2 are not their work but extrapolations by us of their
projections.

The extrapolated USDA projections of feedgrains and soybeans are much
closer to ours than are Abel's. Abel assumes a major increase in production of corn to produce ethanol for combination with gasoline to produce
gasohol. He also anticipates a significant increase in corn production
for use as a sweetener. Both uses of corn yield a high protein residue
which, with price adjustments for quality differences, competes directly
with soybeans as a feed supplement. Abel's assumptions about the growth
of demand for corn for production of ethanol and sweetener thus accounts
for much of the difference between his extrapolated projections and ours
for feedgrains and soybeans in 2010. The extrapolated USDA projections

Table 2-2. Alternative Projections of U.S. Production of Wheat,

Feedgrains and Soybeans

(million metric tons)

	1978/80	USDA 1990	Abel 2005	2010 Extrapolations of		
				USDA	Abel	RFF
Wheat	57.0	77.1	83	114	88	98
Feedgrains	219.9	282.0	404-435	385	441-476	354
Soybeans	54.0	72.1	64-76	105	66-80	120

Note: Abel projects oilseed production, not soybeans only. Typically, soybeans account for 85-90 percent of oilseed production in the United States. We have assumed that soybeans are 85 percent of Abel's projection for 2005.

Sources: 1978/80 - USDA, January 28, 1981; USDA - from table 2-1, this report; Abel - Martin Abel, "Growth in Demand for U.S. Crop and Animal Productions," in Pierre Crosson, The Cropland Crisis: Myth or Reality (Baltimore, Md., Johns Hopkins University Press for Resources for the Future, 1982); RFF - from table 2-1, this report. See appendix to this chapter for details on derivation.

for feedgrains and soybeans differ from ours in the same way as Abel's, but the amount of the difference is much less. The USDA projections also assume increased production of corn for ethanol production, but evidently on a smaller scale than Abel assumes.

Our projections make no special allowance for use of corn to produce ethanol or sweeteners. Abel's assumption that gasohol production will expand substantially is the main reason for the differences between his projections and ours. Federal legislation passed in 1980 called for 10 billion gallons of ethanol production by 1990 to make gasohol, an enormous increase. Should this occur, even Abel's projection of demand for feedgrains would be low. However, Sanderson (1981) has shown that both the economic costs and budget costs (gasohol is subsidized) of achieving the legislated goal would be enormous. He argues, moreover, that by the 1990's, coal, and possibly wood, will be more economical feedstocks for liquid fuel production than grain. Other evidence points in the same direction. At a meeting of the Bio-Energy World Congress and Exposition, held in Atlanta in April, 1980, most speakers agreed that wood eventually will prove more economical than grain as a source of biomass for energy production (Science, 1980, page 1018).

Abel, in discussion with one of the authors, agreed that there is great uncertainty about the long-run course of ethanol production, and that by 2010 wood, or more likely coal, may well replace corn as the main nontraditional feedstock for production of liquid fuel. Abel believes, however, that the corn based production capacity built in the 1980s will be maintained. We are not persuaded on this latter point. By the winter of 1981/82 much of the ardor for gasohol evident in mid-1980 appeared to have cooled. Informal reports received by the authors indicated that some of the plant expansion plans announced in 1980 were being cut back or deferred. The Reagan administration was taking a more hard-eyed look than its predecessor at the subsidies in the gasohol program. All of this, combined with the prospect that by the 1990s alternative feedstocks will prove more economical than grain, suggests to us that by 2010 the demand for grain for ethanol will be insignificant. This is the justification for our decision not to make special allowance for corn-based production of ethanol in our projections of feedgrain production. It follows that we would not make a corresponding adjustment in our projections of soybean production.

With allowance for the difference with respect to treatment of the role of corn in ethanol production, we believe that Table 2-2 indicates that our projections are generally consistent with those of the USDA and Abel. The differences in the total tonnage of grains and soybeans are small for thirty year projections. This, of course, does not mean that we are right, but it does suggest that our thinking about the long-run prospects for U.S. production of grains and soybeans is not grossly different from that of others who have labored in this field. In work of such great uncertainty only gross differences should be taken as significant.[1]

---

[1]Since this was written the National Agricultural Lands Study (1981) published a set of USDA projections for 2000 which imply substantially higher rates of growth in production and exports than any of those shown in Table 2-1. Specifically, the NALS median projections (a range

The RFF projections in table 2-1 are based on no explicit assumptions about crop prices except that their influence on the growth of domestic and foreign demand will be small relative to the influence of world population and income growth. In fact, we think real prices received by American farmers for grain, soybeans, and cotton are more likely to rise over the next several decades than they are to remain at 1978/80 levels. The reason is our expectation that real prices of key inputs--energy, fertilizer, water and land--are likely to increase more than productivity. The basis for these expectations is given in subsequent chapters.

The RFF projections of exports assume that U.S. shares of world trade in grains and soybeans in 2010 will be the same as in the late 1970s. Is this consistent with rising real prices of these crops? The answer depends on the price elasticity of demand for U.S. exports. We have not explored this issue. However, the USDA's NIRAP model, which incorporates crop prices and price elasticities of domestic and foreign demand, shows that the USDA projections given in our table 2-1 are associated with real price increases of 25 to 30 percent for feedgrains and soybeans from 1979 to 1990. Real wheat prices decline, however. Since our production projections conform reasonably well with those of the USDA, we take these NIRAP price results as evidence that our projections are consistent with

---

is given) are 2.7 percent annually for total production from 1980 to 2000 and 5.6 percent annually for total exports. The RFF 2010 projections in table 2-1 imply the following annual growth rates from 1980:

	Production (%)	Exports (%)
Wheat	1.4	1.8
Feedgrains	1.9	2.7
Soybeans	3.0	3.2
Cotton	1.5	not projected

In the NALS demand is projected to grow more rapidly for grains and soybeans than for other components of agricultural production. Consequently, the differences between the RFF projections of those commodities and the projections of them embodied (but not shown separately) in the NALS projections are greater than the above percentages indicate. The NALS projections assume constant relative prices and ours allow for some real increase in prices, as noted below. Nonetheless, it is likely that the constant price equivalents of the RFF projections would be less than the NALS projections. By that comparison, therefore, the RFF projections probably are conservative.

real price increases on the order of 25 to 30 percent, at least for feed-grains and soybeans.

Martin Abel's projections, shown in table 2-2, explicitly assume no change in real crop prices between 1980 and 2005. If real prices in fact change, as the NIRAP projections indicate and we think likely, Abel's projections would be modified accordingly. He asserts (Abel, 1982, page 1) as follows:

> Our projections for the U.S. and the world would have to be modified if alternative real price assumptions were used, and the modifications would depend upon whether the price change was due to factors affecting either demand or supply.

In summary, the projections we use to examine the resource and environmental implications of increased agricultural production in the United States are consistent with rising real crop prices. But the projections assume that the effect on demand of any price increases will be small relative to the effects of world population and income growth, and that, in particular, such increases will not significantly alter the U.S. share of world trade in wheat, feedgrains, and oilmeal.

References

Abel, Martin. 1982. "Growth in Demand for U.S. Crop and Animal Production by 2005," in Pierre Crosson, The Cropland Crisis: Myth or Reality (Baltimore, Md., Johns Hopkins University Press for Resources for the Future).

National Agricultural Lands Study. 1981. Final Report. A joint effort by the U.S. Department of Agriculture, Council on Environmental Quality and ten other federal government agencies (Washington, D.C., GPO).

Sanderson, Fred. 1981. "Benefits and Costs of the U.S. Gasohol Program," Resources no. 67 (July), Resources for the Future, Washington, D.C.

Science. 1980. vol. 208, May 30.

USDA. 1981. Foreign Agricultural Circular Grains, FG-4-81, January 28, Washington, D.C.

_____. 1981. Agricultural Outlook (March).

Chapter 3

FACTORS AFFECTING FARMERS' CHOICES AMONG
TECHNOLOGIES: PRICES OF INPUTS

Introduction

Meeting our projected levels of production of grains, soybeans and
cotton will increase the American farmers' demand for resources. The
amount of the increase for each type of resource depends upon the kinds
of technologies farmers choose. We assume that in making these choices,
farmers seek to maximize their incomes, and that the choices are conditioned fundamentally by the prices and productivities of alternative
technologies. In this chapter, we consider prices of certain key inputs.
Productivity is considered in the next chapter. Farmers' choices also
are affected by public policies. In most cases these will be reflected
in prices of technologies. In some, however, it will not, as when the
Environmental Protection Agency (EPA) bans an insecticide. In these
cases, policy is a separate factor influencing farmers' choices, and we
treat it as such where appropriate.

Types of Technology

Technologies frequently are labeled according to the intensity of
use of some particular input relative to other inputs. For example, a
technology using much labor relative to capital is called labor-intensive,
or labor-using, to distinguish it from another technology in which the
ratio of labor to capital is lower. Similarly, technological innovations
may be described as "saving" or "using" with respect to some particular
input, depending upon whether the proportion of that input to other inputs is less or more in the new technology than in the one it replaces.

We find it useful to distinguish agricultural technologies as land-using or land-saving, according to whether the ratio of land to other inputs is high or low. The implication is that technologies should be thought of are lying along a spectrum ranging from very land-using (high ratio of land to other inputs) to very land saving (low ratio of land to other inputs). A given technology will be land-saving relative to some technologies and land-using relative to others.

Although the distinction between the two kinds of technology lacks precision, it is useful for analysis of the land use and environmental implications of expanding agricultural production. To meet a given increase in production, land-using technologies imply greater demand for land, hence lower growth of yields, than land-saving technologies. The distinction is useful also because the environmental implications of the two kinds of technology are different. With land-using technologies, problems of erosion typically will be more important relative to problems of agricultural chemicals than with land-saving technologies.

As with any analytical device, the distinction between land-using and land-saving technologies will not catch all the nuances of reality. For example, American farmers who expand crop production with land-using technologies typically will also apply fertilizers and pesticides to the land. Consequently, the land-using mode of expansion may entail environmental costs of fertilizer and pesticide pollution as well as costs of erosion. In the same sense, farmers who adopt the land-saving mode may nonetheless cause increased erosion by abandoning erosion control practices, such as strip-cropping, in an effort to maximize production per unit of land. Since either mode of expansion may entail environmental costs associated primarily with the other, the two categories of technology should be regarded as polar cases, the actual technologies employed being chosen from the spectrum in between.

Choices Among Technologies: A Historical Perspective

Table 3-1 indicates that between 1951 and 1972, the trend of American farmers' choices among technologies was weighted toward the land-saving end of the spectrum. In this period, the ratio of all purchased inputs

Table 3-1. Ratios of Inputs to Cropland Used for Crops in U.S. Agriculture
(1967 = 100)

Year	Fertilizer[a]	All purchased inputs
1951–55	35	67
1956–60	46	77
1961–65	70	92
1966–70	108	102
1971	123	105
1972	128	108
1973	123	105
1974	129	102
1975	115	99
1976	136	106
1977	141	109
1978	134	116
1979	141	117
1980	143	111

Source: U.S. Department of Agriculture, Economic Indicators of the
Farm Sector: Production and Efficiency Statistics, 1980, Statistical
Bulletin 679 (Washington, D.C., Economic Research Service, January 1982).

[a]Per acre of cropland harvested.

(not including labor) to cropland used for crops rose 61 percent. The increase in the ratio of fertilizer to cropland was particularly high-- 266 percent.

This trend toward land-saving technologies from the beginning of the 1950s to 1972 was consistent with the behavior of relative input prices in this period. Table 3-2 supports this. The prices of all purchased inputs and of fertilizers fell relative to the price of land, measuring the latter by the per acre value of farmland and buildings (Land$_A$ in table 3-2). The prices of purchased inputs and of fertilizer rose from 1951/55 to 1956/60 relative to the capitalized value of land rents (Land$_B$), but fell thereafter to the early 1970s. Relative to the Land$_A$ price, purchased input prices and prices of fertilizer continued to fall after the early 1970s, but at a much slower annual rate than in the two previous decades. By contrast, purchased input prices and fertilizer prices rose relative to the capitalized value of land rents after 1974, reversing the previous trend. The behavior of input prices relative to land prices since the early 1970s, thus, is consistent with the slower shift to land-saving technologies in that period, the price effects being particularly marked when price is measured by the capitalized value of land rents.

The implications for farmers' choices among technologies of the price ratios in table 3-2 are not as unambiguous as they may seem. The reason is that land prices are themselves affected by farmers' choices but prices of the other inputs are not, at least not in the long run. Prices of fertilizer and other farm inputs are determined primarily by the costs of producing these inputs, or in any case by factors other than farmers' demand for them.[1]

This is not true of land prices, however. The greater part of the demand for farm land comes from farmers.[2] There is interdependence,

---

[1]This probably is less true of fertilizer prices than of other inputs. The United States accounts for 20-25 percent of world consumption of nitrogen and potash fertilizers, and for about 20 percent of consumption of phosphorus fertilizers. These amounts suggest that short-term shifts in U.S. demand likely would affect fertilizer prices.

[2]Duncan (June 1977) indicates that in 1975, 63 percent of farm real estate transfers in the United States were to farmers, about the same as in the previous three decades.

Table 3-2. Ratios of Input Prices to Prices of Land and Prices of
Fertilizer (1967=100)

| | Relative to land | | | | Relative to fertilizer |
| | $Land_A$ | | $Land_B$ | | |
Year	All purchased inputs	Fertilizer	All purchased inputs	Fertilizer	All purchased inputs
1951-55	180	207	181	203	89
1956-60	147	167	294	322	91
1961-65	122	133	223	237	94
1966-70	95	87	159	145	110
1971	91	73	177	143	124
1972	89	69	190	148	128
1973	91	64	103	72	143
1974	85	85	57	57	100
1975	83	98	71	85	84
1976	76	73	94	90	104
1977	69	63	129	117	110
1978	68	56	163	136	120
1979	68	54	na	na	127

Source: $Land_A$, all purchased inputs and fertilizer from USDA, Agricultural Statistics, 1972 and 1980. $Land_B$ from Emery Castle and Irving Hoch, "Farm Real Estate Price Components," American Journal of Agricultural Economics vol. 64, no. 1. $Land_A$ is an index of the value per acre of farmland and buildings. $Land_B$ is the capitalized value of annual net rents earned by farmland, calculated by Castle and Hoch from USDA data.

therefore, between the price of land and farmers' choices between land-using and land-saving technologies. This is not true (or not true to nearly the same extent) for nonland inputs, for which farmers are price takers.

The implication is that movements in the price of farm land reflect movements in the demand and supply curves of land relative to one another. For this reason, our analysis of the role of land in farmers' future choices between land-saving and land-using technologies focuses on factors affecting the demand curve for land and the resulting supply response. Drawing on the analysis in this and the following chapter of the future behavior of the prices and productivity of non-land inputs, we project the amount of land needed to produce our projected amounts of crop output. In a subsequent chapter, we examine the supply response of cropland and make a judgment about the effect on land price of the supply-demand interaction.

In this chapter, we focus on the prices of fertilizer and energy, and on the cost and availability of water for irrigation, because these are the principal land-saving inputs.

Prices of Inputs

Energy

American farmers use large amounts of energy to drive tractors and other farm machinery, to pump irrigation water, to dry crops, and in the form of nitrogen fertilizer and pesticides. With respect to farmers' choices among land-using and land-saving technologies, the two most important uses of energy are for producing nitrogen fertilizer and for pumping irrigation water.[3] According to Dvoskin and Heady, about 30 cubic feet of natural gas are used on average to produce one pound of nitrogen fertilizer. This indicates that at prices prevailing in 1979/80, natural gas accounted for a little over 45 percent of the price of anhydrous am-

---

[3] USDA (1976) shows that in 1974 fertilizer accounted for 30.8 percent of energy used in farm operations in the United States and irrigation for 12.9 percent. Field operations took 25.9 percent, presumably for tractors and other farm machinery.

Table 3-3. Index of Real Prices Paid by Farmers for Fuel and Energy in
the United States (1967=100)

Year	Index	Year	Index
1965	104	1972	86
1966	101	1973	87
1967	100	1974	108
1968	97	1975	110
1969	93	1976	110
1970	89	1977	111
1971	88	1978	109
		1979	112

Sources: U.S. Department of Agriculture, Agricultural Statistics,
1980. The "real" price is the index of nominal prices divided by the
Consumer Price Index. The series for fuel and energy prices paid by
farmers begins in 1965.

monia. This number is only illustrative, but it makes the point that the
cost of natural gas has an important influence on the price of nitrogen
fertilizer.

Table 3-3 shows an index of the real prices paid by farmers for fuel
and energy from 1965 to 1979. This price series dates only from 1965, but
there is little doubt that real energy prices paid by farmers declined
more or less steadily from the beginning of the 1950s to the early 1970s.[4]
After 1973, however, these prices rose sharply and, in 1979, were 28 percent above 1973.

The decline in real energy prices from the 1950s to the early 1970s
contributed importantly to the decline in the real price of fertilizers in
this period (noted above) and to the spread of irrigation (detailed later
in this chapter). Should real energy prices continue to rise, the trends
in prices o nitrogen fertilizer and of irrigation are likely to be less

---

[4]In real terms the electricity, gas, fuel oil, and gasoline components
of the Consumer Price Index (CPI) were all lower in the early 1970s than in
1950. Energy prices paid by farmers must have moved in a similar fashion.

favorable to the adoption of land-saving technologies in the future than
they were in the two decades ending in the early 1970s.

There is a consensus that the real price of energy will continue to
rise. This is evident from a major study done at Resources for the Future
of energy alternatives for the United States (Schurr, et. al.). The
study includes a review of studies done by others of energy prospects
for the United States to the end of the century. All of the studies assume that real energy prices will rise from 1975 to 2000. There are differences among the projections in the amount of price increase, but all
agree that real natural gas prices will rise most and coal and electricity prices least. The average rate of increase for all energy sources
is roughly 2 percent per year from 1975 to 2000.

It should be noted that none of these projections of real energy
prices emerges directly from analyses of trends in world energy supply
and demand. Instead they are based on past trends and on informed judgments about the effect on prices of trends in technology, of known re-
serves and new discoveries, OPEC policies, and so on. The authors of each
of the studies obviously believe that the price projections are consistent
with projections of energy demand and supply, but they are not derived
from the analysis of demand and supply. We make this point not to cast
stones at the price projections--we would do them in the same way. Indeed, our conclusion that real crop prices likely will rise over the next
several decades is based on the same approach. Our point rather is that
the energy price projections are based on ad hoc judgments rather than on
systematic analysis of underlying relationships and tested hypotheses
about the determinants of energy prices. The large role of judgment relative to analysis means that faith (or lack of it) weighs more heavily in
acceptance or rejection of the projections than it would if the roles of
judgment and analysis were reversed. We have no basis for questioning
the judgments, but we recognize that we accordingly accept the price projections largely on faith.

We assume that real energy prices paid by American farmers (directly
and indirectly) will rise steadily to the end of the century and beyond
at a rate of some 2 percent annually, that prices of natural gas will
increase substantially faster than 2 percent (without specifying how much

faster), that electricity prices will rise by less than 2 percent annually, and that prices of other energy sources will rise at rates some-
where between those for electricity and natural gas.

Fertilizers[5]

While energy is an important component in the prices of fertilizers,
particularly nitrogen, other factors also are important. It now is generally agreed that the run-up in fertilizer prices in 1974 and 1975 resulted more from rising pressure of demand on supply than from the increase in energy prices in those years. Similarly, the decline in fertilizer prices after 1975 occurred despite continued increases in energy
prices. Thus, while our projections of rising real prices of energy
imply upward pressure on fertilizer prices, particularly for nitrogen,
other factors may offset this pressure. Projections of fertilizer prices
must take these other factors into account as well as trends in energy
prices.

Table 3-4 shows measures of the "real" price of fertilizers in the
United States. Prices are shown relative to the cost-of-living index
(CPI) and relative to the price of corn. The relation to the price of
corn probably carries more weight in farmers' decisions about the use of
fertilizer. The significant point in the table is that since the mid-
1970s the real price of fertilizer, whether measured against the CPI or
the price of corn, has been sharply higher than in 1970.

Projections in a study of demand and supply in the world fertilizer
industry indicate that the growth of world capacity to produce nitrogen
fertilizers would slightly exceed the growth of consumption from 1978 to
1985, but that for phosphate and potash fertilizers, capacity would increase less than consumption (Harris and Harre, 1979). This suggests
that, other things the same, nitrogen prices would be about the same to

---

[5]This section is based primarily on USDA (Dec. 1978b) and on discussions with people at USDA, the World Bank, and the International Fertilizer Development Center at Muscle Shoals, Alabama. The USDA does much work
on projections of quantities of fertilizers supplied and demanded, but
little or nothing on prices, at least for publication. Of the three institutions contacted, only the World Bank currently projects world fertilizer prices, and these are not published.

Table 3-4.   Indexes of Real Prices of Fertilizers in the United States
(1970=100)

Year	Nitrogen[a]		All fertilizer	
	Relative to CPI	Relative to price of corn	Relative to CPI	Relative to price of corn
1970	100	100	100	100
1971	102	131	99	127
1972	100	94	99	91
1973	103	61	101	61
1974	193	108	149	84
1975	256	186	178	129
1976	174	158	143	130
1977	155	160	132	135
1978	133	142	122	121
1979	134	148	119	122

Sources: Nitrogen prices are from U.S. Department of Agriculture,
1979 Fertilizer Situation FS-9 (Washington, D.C., December 1978); and
Agricultural Prices PR-1-3 (80)(Washington, D.C., June 1980); 1977,
1978, and 1979 are averages of prices of May 15 and October 15 in each
year; consumer price index, corn prices and index of all fertilizer prices from USDA, Agricultural Statistics, 1980.

[a]Anhydrous ammonia, which accounts for almost 60 percent of nitrogen
fertilizer production in the United States.

slightly weaker in 1985 compared to 1978, but that prices of phosphate and potash fertilizers would be higher. "Other things," however, are not likely to be the same. The anticipated increase in the real price of natural gas would tend to increase the cost of production of nitrogen fertilizers, quite apart from trends in consumption and capacity in that industry. Studies done at the World Bank conclude that rising costs of construction of new plants also likely will put upward pressure on real prices of nitrogen fertilizer, and also on prices of phosphates and potash. According to these studies, real prices of the fertilizers in 1985 could be 15 to 20 percent above the levels of 1978. While projections beyond 1985 are not available, we assume the increase in energy prices will put steady upward pressure on fertilizer prices, particularly for nitrogen, and that these prices will rise in real terms to the end of the century and beyond.

Water

Irrigation is a highly land-saving technology, substituting a controlled water supply and other inputs for land. However, little of the water used for irrigation in the United States is traded through markets. Most of it either is pumped from aquifers or rivers by the farmers who use it or is provided by publicly funded surface water systems at highly subsidized prices. Hence prices paid by farmers for irrigation water, where they exist at all, bear little relationship to the true scarcity value of the resource. However, where farmers are free to sell their rights to water to nonagricultural users, the water has an opportunity cost to the farmer, the amount depending upon the strength of the non-agricultural demands. Hence, trends in those demands are an indicator of trends in the "price" of water used for irrigation. The other principal indicator is the cost of pumping groundwater.

Table 3-5 gives information about irrigation in the United States in 1977, and table 3-6 shows irrigated and rainfed land, by region, in corn, soybeans, wheat, and cotton in 1950 and 1977. Together the tables show that in 1977 these crops accounted for about 40 percent of the 46.4 million irrigated cropland acres in the 17 western states. However, in the

Table 3-5. 1977 Irrigated and Dryland Agricultural Land Use by Farm Production Region
(millions of acres)

Farm production region	Irrigation			Dryland			Total		
	Cropland	Pasture	Total	Cropland	Pasture	Total	Cropland	Pasture	Total
Northern Plains	10.6	0.1	10.7	83.9	9.4	93.3	94.6	9.5	104.1
Southern Plains	8.6	0.4	9.0	33.6	27.1	60.7	42.2	27.5	69.7
Mountain	15.2	2.0	17.2	27.1	5.5	32.6	42.2	7.4	49.6
Pacific	11.9	1.4	13.3	11.2	2.8	14.0	23.2	4.1	27.3
17 Western states	46.4	3.9	50.2	155.8	44.7	200.5	202.2	48.5	250.7
Northeast	0.4	0	0.4	16.5	5.8	22.3	16.9	5.8	22.7
Appalachian	0.4	0	0.4	20.4	18.5	38.9	20.8	18.5	39.3
Corn Belt	1.1	0	1.1	88.8	25.2	114.0	89.9	25.2	115.1
Lake States	1.0	0	1.0	43.2	6.9	50.1	44.1	6.9	51.0
Southeast	2.4	1.0	3.4	15.1	13.1	28.2	17.5	14.1	31.6
Delta States	4.0	0.1	4.1	17.2	12.6	29.8	21.2	12.7	33.9
31 Eastern states	9.3	1.1	10.4	201.2	82.1	283.3	210.4	83.2	293.6
National total	55.7	5.0	60.7	357.2	127.8	486.1	412.9	132.7	546.9

Note: Minor differences in totals are due to rounding.

Source: U.S. Department of Agriculture, Basic Statistics: 1977 National Resources Inventory, Revised (Washington, D.C., Soil Conservation Service, February 1980).

Table 3-6. Irrigated and Dryland Acreage of Corn, Sorghum, Wheat, and Cotton for 1950 and 1977

(1,000 Harvested Acres)

Crop		1950	1977
Corn:[a]	Irrigated West	46	8,838
	Dryland West	16,933	4,572
	East	55,486	56,596
	U.S. Total	72,465	70,006
Sorghum:[a]	Irrigated West	894	1,847
	Dryland West	9,435	10,774
	East	13	1,444
	U.S. Total	10,342	14,065
Wheat:	Irrigated West	911	3,899
	Dryland West	48,555	49,387
	East	12,141	12,930
	U.S. Total	61,607	66,216
Cotton:	Irrigated West	1,563	3,868
	Dryland West	6,949	5,212
	East	9,331	4,199
	U.S. Total	17,843	13,279

Note: Irrigated production of soybeans is trivial in the West.

Sources: The total acreage estimates for the West and United States in 1977 are from U.S. Department of Agriculture, Agricultural Statistics, 1978. The 1977 irrigated and dryland acreages are based on data from the Agricultural Census of 1974, adjusted by information from the National Resources Inventory (NRI). Details are from Kenneth Frederick, Water for Western Agriculture (Washington, D.C., Resources for the Future), as are the 1950 estimates.

[a] For grain only.

nation as a whole, most of the land in these crops was not irrigated (table 3-6).

Frederick (1982) shows that rising competition for water from non-agricultural users, increasing energy costs of pumping, and declining water tables will increase the real cost of water for irrigation in the west over the next several decades. Frederick expects continued expansion of western irrigation, but at a much slower rate than in the last couple of decades.

In our terms, this prospect for western irrigation is additional reason to believe that in the future farmers will find land-using technologies more attractive than they did in the fifties and sixties. This implies that the already small percentage of irrigated land in the crops of interest to us will shrink even more. Consequently, the prospect for western irrigation is not of major importance for our purposes.

Table 3-5 indicates that, in 1977, one-sixth of the nation's irrigated land was in the thirty-one eastern states, 75 percent of that in the Southeast (Alabama, Florida, Georgia and South Carolina) and the Mississippi Delta (Arkansas, Louisiana and Mississippi). The figures in table 3-5 understate the amount of irrigated land in the Delta because most of the 776,000 acres of irrigated land in Missouri (a Corn Belt state) are in the southeast corner of the state, a region counted as part of the Delta for soil classification purposes, but not for statistical purposes.

The percentage increase in eastern irrigation exceeded that in the West from 1967 to 1977, with most of the increase in the Southeast, Delta (counting Missouri as part of the Delta) and Lake States (Michigan, Minnesota and Wisconsin). This is evident in table 3-7.

The expansion of eastern irrigation outside the Delta has been primarily on droughty soils in the Lake States and Southeast. Because of low moisture retention characteristics of these soils, plants come under moisture stress very quickly, even though in an average year rainfall is ample to support plant growth (Hanson and Pagano, 1980). The presence of these soils also accounts for most of the expansion of irrigation in Indiana, the Corn Belt state, except for Missouri, with the greatest amount of irrigated land.

Table 3-7. Irrigated Land in the Eastern United States, 1967 and 1977

(1,000 acres)

Region	1967	1977
Northeast	440	370
Appalachia	-	414
Corn Belt	206	1,112
Missouri	118	776
Other Corn Belt	88	336
Lake States	129	966
Southeast	1,694	3,424
Delta	3,520	4,014
Total	5,989	10,301

Sources: James Hanson and James Pagano, "Growth and Prospects for Irrigation in the Eastern United States" (Washington, D.C., Resources for the Future, 1980). Basic data taken from the USDA's Conservation Needs Inventory of 1967 and National Resources Inventory of 1977.

Hanson and Pagano conclude that most of the potential for future expansion in eastern irrigation, apart from the Delta, is in the Southeast, particularly in Georgia and Florida. Aside from the droughty soils in those states, their growing season is long enough to permit double-cropping, and irrigation provides a timely water supply. Conservation tillage, of increasing importance in the Southeast, also encourages double-cropping by saving time between the harvest of the first crop and planting of the second.

Despite the prospect for increased irrigation in the Southeast, the impact on the growth of production and yields of the crops of interest to this study is likely to be small. Material presented in a 1977 federal government inter-agency Task Force report on irrigation, cited by Hanson and Pagano (1980) and in Shulstad and coauthors (1980) indicate that there are about 9 million potentially irrigable acres in the East, apart from the Delta, 6 million of which already are irrigated (table 3-7). Even if all of the remaining irrigable but not yet irrigated land were in the

Southeast, and it is not, the potential for expansion would appear to be only on the order of 3 million acres.

Hanson and Pagano, drawing on the 1974 Census of Agriculture, show that in that year only 5.5 percent of the irrigated land in the Southeast was in crops of interest to this study. While corn and soybeans are expected to be much more important in additional irrigation than at present, they are not likely to account for all of the prospective growth.

In contrast, the study by Shulstad and coauthors suggests that expansion of irrigation in the Delta could have a significant impact on production and yields in that region, particularly of soybeans. These authors consider numerous alternative combinations of crop prices, input costs, yields and crop rotations, and most indicate that it would be economical to irrigate all of the some 17 million acres in the Delta judged by the authors to have the physical characteristics consistent with irrigation. In the alternatives considered, the amount of irrigated soybean land varies from 8 to 12 million acres. These estimates suggest that we can expect a substantial expansion of irrigated land in the Delta, much of it in soybeans. The implications of this for yield projections in the Delta are considered in chapter 5.

Conclusions on Prices of Inputs

We conclude that real prices of energy will rise at an average annual rate of about 2 percent from 1980 to the end of the century and beyond, with natural gas prices increasing substantially more than this. This will put upward pressure on prices of nitrogen fertilizer and on the costs of pumping water for irrigation. Higher prices of natural gas combined with rising costs of construction in the fertilizer industry likely will cause real prices of fertilizer to rise by 1985, even though demand does not press hard on capacity. We assume that real nitrogen prices will continue to rise to the end of the century and beyond. Declining water tables in the High Plains will reinforce the effect of higher energy prices on costs of pumping in that region, and throughout the West growing nonagricultural demands for water will increase its cost to agriculture.

The amount of irrigated land in the East, apart from the Mississippi Delta, is expected to increase, particularly in the Southeast, but the impact on trends in yields and production of crops of interest to this study will be small. The expansion of irrigated soybean production in the Delta, however, could be significant.

These increases in real prices of energy, fertilizer and irrigation water will make land-using technologies more attractive to farmers (except perhaps for those in the Southeast and the Delta) than they were in the 1950s and 1960s. However, farmers' choices will be influenced by the prospective productivity of the alternative technologies and by the costs of increasing the supply of cropland. Productivity trends are examined in the next chapter and the supply of land in the one after that.

References

Castle, Emery, and Irving Hoch. 1982. "Farm Real Estate Price Components," American Journal of Agricultural Economics vol. 64, no. 1 (February).

Duncan, Marvin. 1977. "Farm Real Estate – Who Buys It and How," Monthly Review (June), Federal Reserve Bank of Kansas City.

Dvoskin, D. and E. Heady. 1976. U.S. Agricultural Production Under Limited Energy Supplies, High Energy Prices and Expanding Exports Card Report 69 (Ames, Center for Agricultural and Rural Development, Iowa State University).

Executive Office of the President. 1980. Economic Report of the President 1980 (Washington, D.C.).

Frederick, Kenneth. 1982. Water for Western Agriculture (Washington, D.C., Resources for the Future).

Hanson, James, and James Pagano. 1980. "Growth and Prospects for Irrigation in the Eastern United States" (Washington, D.C., Resources for the Future).

Harris, G.T., and E.A. Harre. 1979. World Fertilizer Situation and Outlook, 1978-1985 (Muscle Shoals, Alabama, International Fertilizer Development Center and National Fertilizer Development Center).

Landsberg, Hans, project director. 1979. Energy: The Next Twenty Years. A study sponsored by the Ford Foundation and administered by Resources for the Future (Cambridge, Mass., Ballinger).

Schurr, Sam H., Joel Darmstadter, Harry Perry, William Ramsay, and Milton Russell. 1979. Energy in America's Future: The Choices Before Us (Baltimore, Md., Johns Hopkins University Press for Resources for the Future).

Shulstad, R.M., R.D. May, B.E. Herrington, and J.M. Erstine. 1980. Expansion Potential for Irrigation Within the Mississippi Delta Region (Fayetteville, Department of Agricultural Economics and Rural Sociology, University of Arkansas).

Sloggett, Gordon. 1977. Energy in U.S. Agriculture: Irrigation Pumping, Agricultural Economic Report 376 (Washington, D.C., USDA).

_____. 1979. Energy and U.S. Agriculture: Irrigation Pumping 1974-77 Agricultural Economic Report 436 (Washington, D.C., USDA).

USDA. Various years. Agricultural Statistics (Washington, D.C., GPO).

____. 1976. Energy and U.S. Agriculture: 1974 Data Base (Washington, D.C.).

____. 1977. Agricultural Prices (Washington, D.C., June).

____. 1977. Agricultural Prices (Washington, D.C., October).

____. 1978a. Agricultural Outlook AO-39 (Washington, D.C., December).

____. 1978b. 1979 Fertilizer Situation FS-9 (Washington, D.C., December).

USDA. 1980. Basic Statistics: 1977 National Resources Inventory, Revised February (Washington, D.C., Soil Conservation Service).

____. 1980. Agricultural Outlook AO-54 (Washington, D.C., May).

____. 1980. Agricultural Prices PR-1-5 (80) (Washington, D.C., May).

____. 1980. Agricultural Prices PR-1-3 (80) (Washington, D.C., June).

____. 1982. Economic Indicators of the Farm Sector: Production and
Efficiency Statistics, 1980, Economic Research Service, Stat. Bull.
679 (Washington, D.C., Economic Research Service, January).

U.S. Senate. 1978. Costs of Producing Selected Crops in the United
States--1976, 1977 and Projections for 1978. Prepared by the Eco-
nomics, Statistics and Cooperatives Service of the USDA for the
Senate Committee on Agriculture, Nutrition and Forestry (Washington,
D.C., GPO, March).

Chapter 4

TRENDS IN PRODUCTIVITY
AND CROP YIELDS IN THE UNITED STATES

Introduction

We concluded in the last chapter that real prices of energy, fertilizer, and (for most farmers) irrigation water likely will rise over the
next several decades, in contrast to their movement in the two decades
prior to the early 1970s. This would make land-using technologies look
more attractive in the future than in that earlier period, other things
the same. However, if the productivity of land-saving technologies were
to rise more than in the earlier period, this would tend to offset the
effect of higher input prices on the relative attractiveness of land-
using and land-saving technologies. Should the productivity of land-
saving technologies rise less than in the earlier period this, of course,
would reinforce the effect of input prices in making land-using technologies look more attractive.

We would like to be able to measure past trends in the productivity
of land-using and land-saving technologies as a basis for projecting
these trends into the future. These projections, in combination with the
projections of input prices discussed in the previous chapter, would provide a basis for judging how farmers in the future might choose between
the two types of technology.

We posit a situation in which the "typical" farmer, faced with rising demand for crops, knowing present prices of inputs, and having expectations about their future prices, asks himself the question, what mix of
land and other inputs should I choose in response to higher demand? We
assume that the desire to maximize profits will be the dominant criterion
of choice. Consequently, the farmer will tend to choose that combination

of inputs which, given input prices, yields the highest output per unit of total input, i.e., that combination with the highest total productivity.

This suggests that examination of trends in total productivity would provide insights about farmers' future choices among technologies. We have not undertaken such an examination, however, because we have serious reservations about the accuracy of the USDA's index of total productivity, especially in the period since 1972. The argument underlying these reservations is given in Appendix B to this chapter. Suffice it to say here that the sharp decline in the last several decades in the quantity and increase in price of farm labor relative to the quantities and prices of other inputs, particularly fertilizer and farm machinery, creates a fundamental problem in constructing an unambiguous index of total input, hence of movements in total productivity. In addition, it appears to us that the way land is handled in the USDA's index of total productivity significantly underestimates the role of land in expanding production after 1972, causing the index to overestimate the increase in total productivity.[1]

For these reasons we have focused our attention on crop yields instead of on total productivity. This has several advantages. One is that our interest in this study is in the resource and environmental impacts of crop production, not total production. Another is that examination of trends in yields provides a basis for projecting them, and therefore of converting our projections of production into projections of land used for crops. A disadvantage of focusing on yields is that they provide only a partial measure of productivity, but farmers' choices are conditioned by total productivity. However, in attempting to explain the behavior of yields we necessarily deal with trends in the productivity of fertilizers and other components of land-saving technologies. This provides insights to movements in total productivity even though we do not measure it directly.

---

[1]For a discussion of the index number problems, and others affecting the USDA's measure of total productivity change, see USDA (February 1980).

Trends in Yields: All Crops

Table 4-1 gives information about actual and trend yields of all
crops between 1960 and 1980. The trend values are based on actual yields
in 1950-1972. Column (3) of the table indicates that in each year after
1972 actual yields were less than trend yields established in the period
1950-1972. Not only did actual yields fall short of trend yields in
every year 1972-1980, but the average annual deviation from trend in
that period was 4.3 times as large as in 1950-1972. In no period from
1950-1972 did actual yields fall short of trend yields for as many as
eight years in a row. (The maximum was four years, 1954-1959.) The
yield experience in 1973-1980, therefore, suggests that the trend of
yields in those years was lower than in 1950-1972.

From table 3-1 in the previous chapter it can be calculated that the
ratio of all purchased inputs to cropland used for crops increased at an
annual rate of 2.5 percent from 1951-55 to 1972, then rose only .3 percent annually from 1972 to 1980. The quantity of fertilizer per acre of
harvested cropland increased 7 percent annually from 1951-1955 to 1972,
but by only 1.4 percent per year from 1972 to 1980. The amount of harvested ,cropland declined from an average of 338 million acres in 1951-
1955 to 289 million acres in 1972, then rose to 346 million acres in
1980 (USDA, January 1982).

Given these changes after 1972 in behavior of ratios of nonland
inputs to land and in the amount of harvested cropland, the slower growth
of crop yields after 1972 is not surprising. Indeed it is the expected
result of the reduced shift to land-saving technologies. With respect to
post-1972 experience the fact that some of the land brought under crops
after 1972 was of lower natural fertility than land already in crops also
contributed to slower growth of yields (more on this in Appendix A to
this chapter).

For these reasons, the slower growth of crop yields after 1972 does
not necessarily indicate that the slower shift to land-saving technologies occurred because the productivity of these technologies, as measured
by yields, had declined, or was rising less rapidly. At least some, and

Table 4-1. Yields for All Crops (1967 = 100)

Year	(1) Actual yields	(2) Trend of actual yields	(3) Actual yields/ trend yields	(4) "Required" yields	(5) "Required" yields/ trend yields
1960	88	88	1.00		
61	92	90	1.02		
62	97	92	1.05		
63	99	95	1.04		
64	96	97	.99		
1965	102	99	1.03		
66	99	101	.98		
67	100	103	.97		
68	105	106	.99		
69	110	108	1.02		
1970	104	110	.95		
71	112	112	1.00		
72	118	114	1.04		
73	113	117	.97	124	1.06
74	103	119	.87	115	.97
1975	110	121	.91	126	1.04
76	110	123	.89	126	1.02
77	115	126	.91	136	1.08
78	119	128	.93	137	1.07
79	126	130	.97	150	1.15
1980	114	132	.86	137	1.04

Note: Column (1): Index of crop production divided by acres of cropland harvested, 1967 = 100.

Column (2): Arithmetic trend of actual yields 1950-1972, projected 1973-1980.

Column (4): Yields that would have been "required" to produce
each year's output 1973-1980 if the amount of cropland harvested had remained the same as in 1972.

Source: U.S. Department of Agriculture, Economic Indicators of the
Farm Sector: Production and Efficiency Statistics, 1980 Economic Research
Service Statistical Bulletin no. 69 (Washington, D.C., January 1982).

perhaps all of the yield behavior was a result of the slower shift, not a cause of it.

To provide some insights into this cause-and-effect relation between yield behavior and the behavior of technology after 1972, we constructed an index of "required" yields for 1973-1980. We asked the question, What levels of yield would have been required if farmers had produced the actual amounts of crops in those years with the same amount of cropland harvested as in 1972? Those yields are in column (4) of table 4-1. We then asked the question, Given the yield experience of 1950-1972, could farmers reasonably have expected to achieve the yields shown in column (4)? We believe column (5) indicates that in most years after 1973 farmers' responses would have been negative; that is, the "required" yields would have appeared to be out of reach, even if farmers had expected the pre-1972 trend in yields to continue. To achieve the higher levels of production after 1972, therefore, farmers would have concluded that more land would have to be brought into production.

This argument indicates that the productivity of land-saving technologies, as measured by crop yields, was not rising fast enough before 1972 to meet the demand for crop production in 1973-1979 without an increase in the amount of land in production. The argument does not indicate, however, that the rate of increase of productivity of these technologies was declining.

These conclusions are useful for our purposes, but they do not lead as far as we would like because they deal with crop production as a whole. Our interest is in specific crops, or groups of crops: wheat, feedgrains, soybeans, and cotton. Examination of the yield experience of these crops is the next step.

Yields of Grain and Soybeans

The analysis is focused on the behavior of yields of principal grains and soybeans in the United States from the late 1940s to the late 1970s.[2]

---

[2]We also are interested in production and yields of cotton. However, there has been no trend in cotton yields since the mid-1950s, and we anticipate no dramatic change in this. Moreover, even if there were such a change, the amount of land in cotton would remain small relative to land

The effort to measure the effect of trends in technology on yields of these crops is complicated by the fact that factors other than technology affect yields. We identify two broad categories of factors. One we call pure shift factors and the other pure yield factors. Yields for a given crop differ among states. Should the relative distribution among states of land in that crop change over time, national average yields will change even if yields in each state remain the same. We say this change in national average yields is attributable to pure shift factors. Changes over time in yields within each state reflect pure yield factors.

We distinguish three pure yield factors: (1) weather, (2) land quality, and (3) technology, by which we mean everything affecting yields within states except weather and land quality.[3] We consider two components of technology: (a) fertilizers; and (b) irrigation.

We have made quantitative estimates of the contributions of pure shift and pure yield factors to the behavior of corn, wheat, sorghum, and soybeans for selected periods from 1946-50 to 1975-76. Within each set of pure yield factors, we have made estimates of the effect of weather and of land quality on yields of corn, wheat, and soybeans over these same periods in principal producing areas. We have not estimated the yield effects of fertilizer on these crops, and our estimates of irrigation effects are incomplete. There is information available, however, about both fertilizer and irrigation which permits some judgments to be drawn about the effects of these inputs on trends in yields.

The analysis, which is quite detailed and technical, is in Appendix A to this chapter. Here we present only a summary of the principal conclusions.

Our analysis shows that pure yield factors accounted for almost all the growth of national yields of wheat and sorghum from 1946-50 to 1975-76,

---

in grains and soybeans. For these reasons we do not consider trends in yields of cotton.

[3] In this analysis the land quality effect shows up when crop production spreads to inferior land. The analysis does not include the effects of erosion over time on a given piece of land. Despite much discussion in the soil conservation literature of the effects of erosion on yields, there are no reliable estimates of these effects on the trend in national, or state, average yields.

and for about 90 percent and 80 percent, respectively, of the growth of corn and soybean yields. The relatively low percentage for soybeans reflects the increasing share of soybean acreage in the Mississippi Delta, where yields are relatively low.

We examined pure yield factors in the growth of corn and soybean yields in the Corn Belt and wheat yields in the Plains states. Specifically, we sought to isolate the effects on yields of (1) weather, (2) land quality and (3) technology. We particularly consider yield behavior before and after 1972 because of the slower growth of yields after that date.

For soybeans we conclude that weather alone could account for post-1972 yield behavior. That is, there was no evidence that the trend of soybean yields after 1972 was affected by changes in land quality or in technology.

In every year from 1973 to 1980 wheat yields in the Plains states were less than weather adjusted trend yields (based on 1950/72). Weather might account for some of this and bringing in additional land of inferior quality probably accounts for more. However, our analysis indicates that the combined effect of weather and additional land does not explain the failure of actual wheat yields to grow in accordance with the pre-1972 trend. So we examined two important components of wheat production technology, fertilizer and irrigation. Per acre applications of fertilizer to wheat land in the Plains states grew more slowly after 1972 than before, which would tend to slow the growth of yields. However, a substantial amount of wheat land receives no fertilizer so the slower growth in fertilizer use is not likely to have been of major importance. Agricultural census data do not show that the expansion of irrigated wheat land in the Plains states from 1974 to 1978 was significantly slower than in the 15 years prior to 1974.

In short, it appears that components of wheat production technology other than fertilizer and irrigation must account for the relatively slow growth of wheat yields after 1972 in the Plains states. This indicates an important line of research, but we were unable to pursue it in this study.

Actual corn yields in the Corn Belt fell short of the trend values
of weather adjusted yields in 6 of the 8 years from 1973 to 1980. Our
analysis indicates, however, that the combined effect of weather and
incorporation of inferior land could easily explain this. The evidence,
therefore, does not indicate that technology contributed less to the
growth of corn yields in the Corn Belt after 1972 than it did in the two
preceding decades.

National level data show that per acre applications of fertilizer
to land in corn, wheat and soybeans increased much more slowly after
1972 than previously. The effect of this on the growth of national
yields of these crops is not clear, however. As just noted, much land
in wheat receives no fertilizer, and this is true also of soybeans. The
yield effects of changes in the amount of fertilizer applied to these
crops, therefore, could easily be obscured by other factors. Most corn
land is fertilized, but slower growth in the application rate does not
appear to have affected the growth of yields, at least not in the Corn
Belt. It is possible that the increase in fertilizer prices after 1973
induced farmers to apply it more carefully to avoid waste. In this case,
the slower growth in per acre application of fertilizer would not slow
the growth of nutrient availability to plants, at least not in the same
proportion.

At the national level, the expansion of irrigated production of wheat
and corn from 1950 to 1977 was large in percentage terms, and yields of
irrigated land in these crops rose. However, as noted in the last chapter, irrigated land in these crops was small in relation to dryland, and
dryland yields also rose sharply. Consequently, the contribution of
irrigation to the growth of national yields of corn and wheat was small.

Conclusions About Trends in Technology

The analysis of trends in all crop yields and in yields of grains
and soybeans does not demonstrate that the slower shift toward land-
saving technologies after 1972 occurred because the productivity growth
of these technologies was slowing. The slower growth of per-acre fertilizer use on wheat, corn, soybeans and other crops after that date does

not necessarily reflect diminishing marginal productivity of fertilizers. It could as well be explained by the increase in the real price of fertilizer.

There is nothing in our analysis, however, to suggest an upward shift in the trend of productivity of land-saving technologies. In the absence of such a shift, our projected rise in real prices of energy, fertilizer, and irrigation water implies that the slower shift toward land-saving technologies that set in after 1972 will continue over the indefinite future. In this event, our projections of increased production mean that the demand for cropland will continue to rise, suggesting two questions: how much might the demand for land increase, and what will be the supply response? We examine these questions in the next chapter.

References

USDA. 1980. Measurement of U.S. Agricultural Productivity: A Review of Current Statistics and Proposals for Change, Economic Research Service Technical Bulletin 1615 (Washington, D.C., February).

____. 1982. Economic Indicators of the Farm Sector: Production and Efficiency Statistics, 1980, Economic Research Service Statistical Bulletin no. 679 (Washington, D.C., January).

Chapter 5

THE DEMAND FOR AND SUPPLY OF CROPLAND

Introduction

We have argued that prospects for the prices and productivity of
land-saving inputs suggest that farmers will continue to select technologies from the land-using end of the spectrum. If this happens, how
much land will American farmers need to produce the levels of crop output
projected in chapter 2? This is the first of two principal questions
addressed in this chapter. The second question is in two parts:
(a) will the present supply of cropland be adequate to meet the projected
demand? (b) If present supply is not adequate, what will be the effect
on the economic cost of agricultural land of converting non-cropland to
crops?

We proceed in three steps: (1) projection of regional production
of each crop; (2) projection of regional yields; (3) analysis of the
present and potential supply of land, by region, in relation to projected
demand. A detailed account of these steps is in the appendix to this
chapter. Here we give a summary of the underlying arguments.

Regional Projections of Production

These projections are derived from projections of regional shares
of national production. Table 5-1 shows actual shares for various years
and projected shares for 1985 and 2010. The table indicates that for
the past years shown the regional shares of wheat, feed grains, and soybeans were quite stable in the sense that the absolute changes in percentages generally were small. The maximum change was 7.5 percentage
points in the share of the Northern Plains in wheat production between

Table 5-1. Percentage of Actual and Projected Regional Shares of Crop Production

Crop	North-east	Lake States	Corn-Belt	Northern Plains	Appa-lachia	South-east	Delta	Southern Plains	Moun-tain	Pacific
Wheat										
1967-69	1.9	4.3	12.1	40.5	1.9	.7	1.7	11.7	14.2	9.9
1970-72	1.3	4.3	10.2	45.2	2.1	.8	.9	8.7	15.3	11.2
1973-75	1.3	6.2	10.2	39.8	1.9	.6	.8	13.1	14.9	11.3
1976-78	1.1	7.5	11.1	37.7	1.6	.4	1.2	12.5	15.4	11.5
2010	1.0	6.0	5.0	34.0	1.0	1.0	7.0	16.0	16.0	13.0
Feedgrains										
1960-64	3	13	46	17	5	3	1	6	3	3
1965-69	3	13	48	17	4	3	1	6	3	2
1970-74	3	13	46	18	4	3	a	7	3	2
1975-78	3	13	47	18	4	3	a	6	3	2
2010	3	15	46	15	5	4	0	6	4	2
Soybeans										
1967-69	1.1	8.2	57.6	4.8	6.6	4.1	16.4	1.1	--	--
1970-72	1.0	7.9	59.7	3.9	6.8	4.5	15.5	.6	--	--
1973-75	1.2	8.5	56.8	4.7	7.8	5.9	14.2	.9	--	--
1976-78	1.3	8.4	56.2	4.2	7.8	5.7	14.9	1.4	--	--
2010	1.0	8.0	48.0	5.0	8.0	11.0	16.0	4.0	--	--
Cotton										
1967-69	--	--	2.0	--	4.6	9.4	28.1	34.7	8.1	13.1
1970-72	--	--	3.1	--	5.7	10.7	31.9	30.9	6.2	11.6
1973-75	--	--	1.9	--	4.0	8.9	25.8	32.0	8.2	19.2
1976-78	--	--	1.6	--	2.5	4.7	23.3	38.0	7.8	20.1
2010	--	--	1.0	--	2.0	3.0	14.0	47.0	8.0	25.0

51

Note: The states in each region are as follows: Northeast--all of New England plus New York, New Jersey, Pennsylvania, Delaware, and Maryland; Lake States--Michigan, Minnesota, and Wisconsin; Corn Belt--Ohio, Indiana, Illinois, Iowa, and Missouri; Northern Plains--North Dakota, South Dakota, Nebraska, and Kansas; Appalachia--Virginia, West Virginia, Kentucky, Tennessee, and North Carolina; Southeast--South Carolina, Georgia, Florida, and Alabama; Delta--Arkansas, Mississippi, and Louisiana; Southern Plains--Oklahoma and Texas; Mountain--New Mexico, Arizona, Nevada, Colorado, Wyoming, Montana, Idaho, and Utah; Pacific--California, Oregon, and Washington.

May not add to 100 because of rounding.

Source: U.S. Department of Agriculture, Agricultural Statistics, various years. Note that the years shown for feedgrains are different from those for wheat, soybeans, and cotton. This reflects the way the data were compiled, and has no significance for the projections.

[a]Less than .5 percent.

1970-72 and 1976-78. This decline was almost exactly offset by an increase in the shares of the Southern Plains and Lake States in wheat production.

A check of data for the 1950s, not shown here, indicates that regional shares in those years were comparable to those shown in table 5-1, with a couple of exceptions. In the 1950s the Corn Belt had 14 to 15 percent of wheat production, compared to the 10-12 percent shown in table 5-1. Most of the decline in the Corn Belt's share of wheat production since the 1950s was offset by increases in the shares of the Southern Plains and Pacific states.

The other significant change since the 1950s was a decline in the Corn Belt's share of soybean production from about 68 percent to the 56 to 60 percent shown in table 5-1. Most of this was offset by an increase in the Delta's share of soybean production from about 9 percent in the 1950s to the 14 to 16 percent shown in the table and by an increase in Appalachia's share from about 5 percent to 7 to 8 percent.

The distinctive feature about the regional distribution of cotton production is its shift from the Southeast and Delta to the Southern Plains (mostly Texas); Mountain states (mostly Arizona); and Pacific (California). According to Collins, Evans, and Barry (1979, p. 21) this shift reflects better conditions for growing cotton in the west than in the Southeast and Delta. In particular, fewer pest problems and better water control through irrigation have led to lower production costs in Texas, Arizona, and California than in the two more eastern regions.

The stability of regional shares in production of grains and soybeans suggests that radical shifts in the regional location of production of these crops are unlikely over the next several decades.[1] We do expect some changes, however. These are: (1) an increase in the share of the Southeast in corn and soybean production and of the Mississippi Delta in soybean and wheat production. In both cases the reasons for expecting increased shares are the potential for supplemental irrigation, discussed

---

[1]The effect of higher energy prices on transportation costs may cause some shifts in the regional distribution of production not captured in our projections. We were unable to explore this possibility in this study.

in chapter 3, and the availability of land now in pasture and forest with potential for conversion to cropland. (2) An increase in the share of the Southern Plains in wheat production is expected, continuing a trend evident for some years and consistent with that region's relatively abundant supply of land with potential for conversion to crops. (3) The decline in the share of the Mississippi Delta in cotton production is expected to continue, with the Southern Plains (mainly Texas) and the Pacific (California) gaining. As noted above, this shift has been under way for some time, and reflects lower costs of cotton production in Texas and California. Texas is expected to gain also at the expense of New Mexico and Arizona (the two mountain states that grow cotton), reflecting the increasing cost and scarcity of water for irrigation. Most of the land in cotton in those states is irrigated compared with 40 to 45 percent in Texas. Most cotton acreage in California (the Pacific state that grows cotton) is irrigated, but California has other cost advantages sufficient to maintain its share of production at a relatively high level.

National and Regional Projections of Yields

If we are right in thinking that in the future farmers are likely to continue to favor relatively land-using technologies, as they have since 1972, then the behavior of yields since 1972 should provide a guide for projections of future yields. The trend of yields of the various crops since 1972 is obscured, however, by extreme variations in weather. The weather effects are particularly troublesome because they severely depressed yields early in the period (for the indexes of weather effects for corn and soybeans, see tables 4-2A and 4-3A of appendix A to chapter 4) and elevated them in 1978 and 1979. (The effect of weather was strongly adverse in 1980, however.) This pattern of weather effects gives a sharp upward tilt to the trend of yields of corn and soybeans, and to a lesser extent, of wheat between 1973 and 1980.

Table 5-2 shows national yields of these crops, and of cotton, from 1971 to 1980. Table 5-3 shows average yields for 1975/78 and projections to 1985 and 2010 for the nation and for the 10 USDA producing regions. The 8 years 1973 to 1980 obviously provide an uncertain base from which

Table 5-2. Yields in the United States of Wheat, Feedgrains, Soybeans, and Cotton

Year	Wheat	Feedgrains (metric tons/acre)	Soybeans	Cotton (lb/acre)
1971	.92	1.77	.75	438
1972	.89	1.98	.76	507
1973	.86	1.81	.75	520
1974	.74	1.49	.63	442
1975	.84	1.75	.78	453
1976	.82	1.82	.71	465
1977	.83	1.88	.83	520
1978	.86	2.08	.81	421
1979	.93	2.31	.87	551
1980	.89	1.95	.73	421

Sources: U.S. Department of Agriculture, Agricultural Statistics, 1980, for 1971-1979; and USDA, Agricultural Outlook, AO-65 (Washington, D.C., Economics and Statistics Service, May 1981) for 1980.

to project national and regional yields. Consequently, we saw little justification for projecting trends derived from mathematical fits to the
1973-1980 data. Instead, for each crop we examined, the behavior of national yields and yields for the two or three major producing regions, and
made judgmental projections on that basis, assuming that the trend of
national yields would be dominated by trends in the major regions. Yields
in other regions were then projected proportionately to one another and so
as to be consistent with the projections for the nation and the major
regions.

These obviously are rough-and-ready procedures, and the resulting
projections of yields are subject to an uncomfortably large margin of
error. The main issue with respect to yields, however, is whether their
post-1972 trend behavior is more likely to presage the future than their

Table 5-3. Yields in the United States of Wheat, Feedgrains, and Soybeans (metric tons/acre)

Region	1975/1978			1985			2010		
	Wheat	Feedgrains	Soybeans	Wheat	Feedgrains	Soybeans	Wheat	Feedgrains	Soybeans
Nation	.84	1.88	.78	.91	2.28	.87	1.08	2.94	1.10
Northeast	.94	1.96	.74	1.07	2.17	.80	1.14	2.79	.98
Lake states	.95	1.95	.86	1.03	2.36	.93	1.15	3.04	1.14
Corn Belt	1.07	2.37	.90	1.16	2.87	.99	1.33	3.70	1.30
Northern Plains	.76	1.62	.74	.83	1.96	.80	.92	2.53	.98
Appalachia	.90	1.87	.67	.97	2.10	.72	1.09	2.72	.89
Southeast	--	1.35	.59	--	1.52	.70	a	1.93	.94
Delta	.95	--	.63	1.03	--	.74	a	--	1.00
Southern Plains	.66	1.62	.67	.71	1.80	.72	.81	2.30	.89
Mountain	.82	1.35	--	.89	1.52	--	.99	1.94	--
Pacific	1.23	1.47	--	1.33	1.68	--	1.49	2.29	--

Sources: 1975/1978 from U.S. Department of Agriculture, Agricultural Statistics, various years, 1985 and 2000 projected as described in the text.

[a] Assumes that all wheat in the Southeast and Delta is double-cropped with soybeans.

pre-1972 trends. Believing, as we do, that the prices and productivities
of yield-increasing inputs will behave more like they did after 1972 than
before that date, we have to conclude that the post-1972 trends in yields
are more indicative of the future than pre-1972 trends. If we are correct
in this, our projected yields, despite the large error to which they are
subject, will be closer to actual yields in 1985 and 2010 than if we were
to assume that the pre-1972 trends will reassert themselves.

The yield projections have enormous implications for projections of
land use and resulting economic and environmental pressures on the nation's
land and water resources. The effect of alternative yield projections is
particularly evident by 2010. Given our projections of crop production,
it is not too much to say that yield behavior will determine whether or
not the nation's agriculture will encounter serious economic and environmental problems over the next several decades. We take the yield projections in table 5-3 as a point of departure in our subsequent analysis.
We emphasize here, however, that the results of that analysis, particularly as they bear on the severity of future environmental pressures,
depend crucially on our projections of yields.

Demand for and Supply of Cropland

Demand for Cropland for Crop Production

The demand curve for cropland shifts to the right in response to rising demand for crops and to farmers' choices of the most economical mix of
land and other inputs with which to meet crop demand.[2] We take yields to
reflect the outcome of these choices. Taken together, therefore, projections of crop production and yields define projected demands for cropland.
These are shown in table 5-4. The table shows harvested cropland in main
crops (feedgrains, soybeans, wheat and cotton) and in other crops, as well
as other uses of cropland.[3] We made no explicit projections of the crop

---

[2]We here leave aside other reasons for demanding cropland, e.g., as
an inflation hedge or tax haven.

[3]For a more detailed discussion of agricultural demands for cropland
for other than main crops see the appendix to this chapter.

Table 5-4. Projections of Demand for Cropland in 2010

(millions of acres)

Region	Harvested Main crops	Other crops	Failed	Fallow	Idle	Total
Northeast	5.9	7.1	.2	--	1.0	14.2
Lake states	30.3	12.7	.8	--	2.8	46.6
Corn Belt	92.3	5.1	.7	--	2.9	101.0
Northern Plains	63.8	20.6	3.3	15.9	3.0	106.6
Appalachia	18.6	5.6	.3	--	1.9	26.4
Southeast	22.1	4.4	2.0	--	1.4	29.9
Delta	20.6	3.2	.3	--	1.4	25.5
Southern Plains	42.8	5.5	1.2	.8	3.3	53.6
Mountain	24.7	9.3	1.1	9.5	1.5	46.1
Pacific	13.8	8.7	.2	3.6	.8	27.1
United States	334.9	82.2	10.1	29.8	20.0	477.0

Note: Main crops are wheat, feedgrains, soybeans, and cotton, projected as described in the test and in the appendix to this chapter.
Land in "other crops" is mostly hay (60 percent). The projections for
non-hay "other crops" is the amount of land in these crops in 1977. The
projections for failed and fallow land also are the same as 1977. Idle
land is the same as 1974.

and cropland prices associated with the projections of demand for land.
The price issue is discussed later, however, in the section called
"Matching Demand and Supply."

Nonagricultural Demands for Cropland

The growth of population and nonagricultural economic activities
exerts increasing demand for land, some of which is cropland or potential
cropland. This source of demand, therefore, must be considered.

Surveys done by the USDA's Soil Conservation Service (SCS) indicate
that between 1967 and 1975 about 3 million acres of agricultural land

were converted each year to urban, transportation, and other non-agricultural uses. Only some 675,000 acres per year were cropland, however,
and perhaps 200,000 acres were potential cropland (Brewer and Boxley).[4]
If this rate of conversion continues (875,000 acres per year), then the
cumulative additional demand for cropland and potential cropland for non-
agricultural uses will be some 25-30 million acres between 1977 and 2010.
In fact, however, the annual rate of conversion is likely to be less than
that experienced in 1967/75.[5] One of the principal factors underlying
that experience was urban population growth and its continued spread into
formerly rural areas. While population will continue to grow and to demand additional land in rural areas for residential purposes, the absolute
rate of growth is expected to decline, which should reduce pressure on
the rural land base relative to 1967-75.

Another important source of pressure on the rural land base in 1967-
75 was construction of the interstate highway system, now substantially
completed.

Strip mining of coal likely will become more important over the next
several decades in response to rising demands for energy and diminishing
supplies of petroleum and natural gas. However, the amount of land used
for this purpose is expected to be quite small, and not all of it will
come from land now constituting the present and potential supply of cropland.[6]

Thus the annual rate of conversion to urban and similar uses of present and potential cropland is likely to be less in 1977-2010 than in
1967-75. We have no reliable way of estimating how much less, but total

---

[4]Fischel (1982) makes a strong case that the SCS surveys overstated
the rate of conversion of agricultural land from 1967 to 1975. We take
the estimates from the surveys as upper bounds.

[5]See Brown and Beale (1981) for an analysis of factors determining
future rates of conversion of agricultural land to non-agricultural uses.

[6]Brewer and Boxley (1982) cite a study showing that by 2000 only
about 1.8 million additional acres would be needed for strip-mining and
siting of coal-fired power plants and associated facilities.

conversion over the thirty-three year period of 20-25 million acres is reasonable.[7]

Supply of Cropland and Potential Cropland

Table 5-5 shows estimates of the amount of cropland and potential cropland in the United States in 1977. We have no information that would account for the differences between the ESCS and NRI estimates of cropland but take some comfort from the fact that the difference between the totals is relatively small. We work with the USDA's National Resources Inventory (Feb. 1980) because, in addition to data on the amount of present cropland, it also includes the estimates of potential cropland shown in table 5-5, of amounts of erosion from cropland, and other useful information.

It is important to note that we have neither national nor regional supply curves for cropland, i.e., schedules showing the amounts of such land that would be supplied at various prices of land.[8] The estimates of 1977 cropland represent a point on such a schedule, indicating the amount of land available for crop production (and related uses) under the crop and input prices and other relevant conditions prevailing in the late 1970s. The figure for potential cropland also is a point estimate. The SCS defined land with high potential for conversion to crops as that with "favorable physical characteristics" and which had been converted in the vicinity in the three years preceding the survey (1974-76). Commodity prices and development and production costs of 1976 were assumed. Medium potential land also has favorable physical characteristics, but conversion costs are estimated to be higher than for high potential land. There is no indication in the NRI, however, that the price-cost relationships of 1976

---

[7] Robert Gray, executive director of the federal government's National Agricultural Land Study expects total conversion of 23-25 million acres of cropland and potential cropland from 1980 to 2000, an annual rate of 1.15 to 1.25 million acres (Journal of Soil and Water Conservation, 1981, p. 63). This would exceed the annual rate of conversion of such land in 1967-75, and is too high, in our judgment, for reasons given above.

[8] Amos (1979) and Shulstad and coauthors (1979) develop proper cropland supply curves for Iowa and the Mississippi Delta. These provide useful analytical model and empirical results for the acres studied, but are of limited usefulness for our purposes here.

Table 5-5. Estimates of the Amounts of Cropland and Potential Cropland
in the United States in 1977, by Major Use
(million acres)

Cropland	ESCS[a]	NRI[b]
Row crops and close grown crops	292.1[c]	308.1
All hay	60.7	63.4
Fallow	30.0	29.3
Tree and bush crops and vineyards	3.4	5.5
Idle	20.0[d]	–
Other	–	6.8
Total	406.2	413.2
Potential cropland[e]		125
High potential		36
Medium potential		89

Note: Excludes cropland used only for pasture, which was 83.5
million acres in 1974.

[a]The Economics, Statistics and Cooperative Service of the USDA
(Economic Research Service as of 1981).

[b]The National Resources Inventory, conducted by the Soil Conservation Service of the U.S. Department of Agriculture in February 1980.

[c]Planted acres of all crops except hay, tree and bush crops and
vineyards, from U.S. Department of Agriculture, Crop Production 1979
Annual Summary, CrPr 2-1(80) (Washington, D.C., USDA, January 15, 1980).
Includes land on which crops failed.

[d]From the 1974 Agricultural Census. The figure for 1977 is not
known, but for reasons given in the appendix it was unlikely to have
been less than 20 million acres.

[e]Defined in the text.

uniquely defined the economic conditions for conversion of land to crops, that is, the NRI does not state that had prices been slightly lower or costs slightly higher conversions would not have occurred. The only statement is that, under price-cost conditions prevailing in 1976, land of the type indicated had been converted. Given this ambiguity, it is conceivable that the amount of land converted in 1974-76 would have been about the same even if price-cost conditions had been less favorable than they were. Changes in the amount of land in crops and in crop prices from 1976 and 1979 also suggest caution in interpreting the NRI's definition of the economic conditions underlying the estimates of convertible land. Table 5-6 indicates that changes in real crop prices from 1976 to 1979 would have favored the expansion of land in wheat and cutbacks in land in all the other crops. In fact, land planted to wheat in 1979 was 8.5 million acres less than in 1976, land in corn in 1979 was about the same as in 1976, land in sorghum, oats, and barley was down in aggregate about 7 million acres, and soybean acreage was up 21.4 million acres, an increase of 42.6 percent (USDA, Agricultural Statistics 1979; and June 27, 1980). Total harvested land rose 12 million acres (USDA, January 1982). Much of the additional land must have been converted from other uses. There was little usable idle cropland in 1976, and since wheat and soybeans generally are not grown in the same areas, not much of the land that went out of wheat would have gone into soybeans.

If real crop prices had to be at least as good as they were in 1976 to encourage conversion of land to crops, the actual conversion experience from 1976 to 1979 is inexplicable. To be sure, these are highly aggregated numbers and a more detailed analysis might resolve the apparent contradiction between actual conversions and the experience expected from the NRI's definition of potential cropland. For example, crop yields rose each year from 1976 to 1979, suggesting that productivity improvements may have offset the negative effects of lower crop prices on farmers' incentives to bring in additional cropland. Another possibility is that farmers' expected higher real prices over the longer term, so went ahead with investments in additional land in spite of the unfavorable short term (given these expectations) movement of prices.

Table 5-6.   Prices Received by Farmers for Grains and Soybeans
             ($/bushel)

Crop	1976	1976 in 1979 $	1979
Wheat	$2.73	$3.38	$3.82
Corn	2.15	2.66	2.41
Sorghum	2.03	2.51	2.33
Oats	1.56	1.93	1.36
Barley	2.25	2.78	2.31
Soybeans	6.81	8.42	6.19

Sources: 1976 and 1979 from U.S. Department of Agriculture, Agricultural Statistics, 1979 and 1980, respectively. The GNP deflator was used to convert 1976 prices to 1979 dollars.

So the crop price and land conversion experience from 1976 to 1979 does not indicate that the NRI estimates of potential cropland are wrong. It does suggest, however, that they should be viewed as no more than rough approximations, and perhaps more likely to be low than high.

Matching Demand and Supply

Table 5-4 projects demand for cropland for crop production at 477 million acres in 2010, an increase of 64 million acres from 1977, judging from the NRI's estimate for that year (table 5-5). Adding to this an increase of some 20 to 25 million acres of cropland and potential cropland demanded for urban and other non-agricultural uses, the total projected demand for cropland in 2010 comes to 497 to 502 million acres and the projected increase is 84 to 89 million acres.

Table 5-7 shows national and regional estimates of the supply of cropland and potential cropland in 1977 and projections to 2010 of demand for such land for crop production as a percent of 1977 supply. The increased demand for cropland for non-agricultural uses is shown only for the nation because we had no basis for projecting this for regions.

The table indicates that except in the Northeast projected demand exceeds 1977 supply by amounts varying from 6 percent (the Lake States)

Table 5-7. Actual and Projected Supply of and Demand for Cropland

| Region | Supply of cropland in 1977 (million acres) | | | Demand in 2010 as percentage of 1977 supply cropland | | | |
| | In use | High potential | High and medium potential | In use | In use plus | | Total demand [b] |
					High potential	High and medium potential	
Northeast	16.9	18.0	21.6	84.0	78.9	65.7	
Lake states	44.1	46.4	52.3	105.7	100.4	89.1	
Corn Belt	89.9	94.7	104.3	112.4	106.7	96.8	
Northern Plains	94.6	99.7	112.5	112.7	106.9	94.8	
Appalachia	20.8	25.5	35.1	126.9	103.5	75.2	
Southeast	17.5	22.4	33.3	170.9	133.5	89.8	
Delta	21.2	24.3	31.3	120.2	104.9	81.7	
Southern Plains	42.2	47.4	62.2	127.0	113.1	86.2	
Mountain	42.2	45.4	56.4	109.2	101.5	81.7	
Pacific	23.2	24.8	28.4	116.8	109.3	95.4	
Nation	412.6[a]	448.6	537.1	115.6	106.3	88.8	92.5-93.5

Sources: Present and potential cropland from U.S. Department of Agriculture, Basic Statistics: 1977 National Resources Inventory, Revised (Washington, D.C., Soil Conservation Service, February 1980). Definitions of high and medium potential land are given in the test. Demand for cropland taken from table 5-4.

[a]Excludes 293,000 acres in Hawaii and 363,000 in the Caribbean.

[b]Includes increased demand of 20 to 25 million acres for nonagricultural uses.

to 71 percent (the Southeast). For the nation the excess demand is 16 percent. If the 36 million acres of high potential land are converted to crops (see table 5-5), demand still exceeds supply in every region except the Northeast, although by smaller amounts, of course. Only if some of the medium potential land is converted would supply equal demand in all regions. The regional margins would be somewhat smaller than shown in table 5-7, however, because the regional projections do not include additional demands for cropland for non-agricultural uses.

The projections represent intersections on demand and supply curves for cropland. We have not attempted to project the corresponding real prices for such land, but we think that in most if not all regions they would be higher than in the late 1970s. There are several reasons. One is that the costs of cropping high and medium potential land must have been higher than costs on land in crops in 1977. (Otherwise the potential land would have been in crops too.) Moreover, over half of the additional land would be that classified in the NRI as having medium potential for conversion. Compared to high potential land this is land that requires more preparation, e.g., clearing, draining, leveling, and so on. Conversion costs, therefore, are generally higher for medium potential land than for that with high potential (Brewer and Boxley, 1982), suggesting that costs would tend to rise as more and more medium potential land is brought under crops.

The opportunity costs of potential cropland also are likely to increase. In 1977 that land was in pasture, forest and range. Converting it to crops will reduce the supply available for those purposes. At the national level the total supply of forest and rangeland is large relative to that with potential for crops, so overall the effect of conversion on prices of forest and rangeland likely would be small. In the Southeast and Mississippi Delta, however, the amount of forestland with potential for crops is large enough to suggest that conversion could significantly increase prices of such land. The NRI shows that almost 40 percent of land in pasture in 1977 had crop potential. Since converting pasture to crops generally is less costly than converting forest or range, prices of pasture land likely would be most affected by conversion to cropland. No doubt there is range and forestland that could be converted to pasture land, moderating the tendency of pasture land prices to rise. But pressure on prices of forest and range land then would be higher.

The projected increase in demand for cropland for non-agricultural uses is small in relation to the projection of demand for land for crops. However, the effect of non-agricultural demand on cropland prices likely would be greater than the increase in physical quantity demanded would suggest. Conversion of cropland to non-agricultural uses occurs typically because the value of the land in those uses is higher than it is in crops. Usually, the difference is large so that when non-agricultural demands begin to compete with crop demand, cropland prices rise steeply. This competition, of course, has been going on since the nation's beginnings without producing a steady increase in cropland prices. Castle and Hoch (1982), for example, show that real agricultural land prices measured by the capitalized value of rents earned in agriculture declined from the end of World War II until the early 1960s, even though this was a period of rapid growth in urban demands for land. We have not explored the reasons for this (neither did Castle and Hoch) but a possible explanation is that the demand for land for crop production was declining in this period.[9] Indeed, the decline in demand for this purpose was substantially greater than the increase in demand for non-agricultural uses.

In contrast to the experience from the late 1940s to the early 1960s, future increases in demand for cropland for non-agricultural uses will be accompanied by sharply rising demand for land for crop production (according to our projections). Pressure on cropland prices from rising non-agricultural demands thus will be reinforced, rather than offset, by demands for crop production. Since non-agricultural users typically can afford to pay much more than farmers for land, the increased competition between them could bid up prices out of proportion to the additional physical quantities of land demand by each.

We conclude that if the demand for cropland rises as we have projected, real prices of the land will rise. But will real crop prices provide

---

[9]Cropland harvested declined from an average of 347 million acres in 1946/49 to an average of 291 million acres in 1961/63 (USDA, January 1982).

sufficient incentive to farmers to increase demand for land on the projected scale in the face of rising land (and other) costs? We cannot definitely answer the question since we have not made specific projections either of crop prices or land prices. We noted in chapter 2, however, that our projections of crop production are consistent with real price increases of 25 to 30 percent. The projections of production and demand for additional cropland, therefore, can accommodate increasing costs of land and other inputs. Since we anticipate continued, albeit relatively slow, increases in productivity, the combined increase in the cost of land and other inputs could exceed 25 to 30 percent without depriving farmers of incentive to expand crop production and land use on the projected scale.

Conclusion

We concluded that the projections of cropland demand and supply in table 5-7 fit plausibly in the scenario for American agriculture developed in this and previous chapters. The scenario indicates that trends in crop demand, particularly for export, in real prices of key land-saving inputs, in technology, and in the supply of cropland portend rising real economic costs of agricultural land and of production. But what of environmental costs? The projections of the amount of land in crops developed in this chapter are crucial to answering this question. In addition, however, we need to consider the amounts of fertilizers and pesticides farmers are likely to use with the land to meet projected demand for crops. This is the business of the next chapter.

References

Amos, O. M. 1979. "Supply of Potential Cropland in Iowa," (Ph.D. thesis, Iowa State University, Ames).

Brewer, M., and R. Boxley. 1982. "The Potential Supply of Cropland," in P. Crosson (ed.), The Cropland Crisis: Myth or Reality? (Baltimore, Md., Johns Hopkins University Press for Resources for the Future).

Brown, D. L., and C. L. Beale. 1981. "Sociodemographic Influences on Land Use in Non-Metropolitan America," in U.S. Senate Agricultural Land Availability (Washington, D.C., Committee on Agriculture, Nutrition and Forestry, GPO).

Castle, E., and I. Hoch. 1982. "Farm Real Estate Price Components, 1920-1978," American Journal of Agricultural Economics vol. 64, no. 1 (Feb.).

Collins, K. J., R. B. Evans, and R. D. Barry. 1979. World Cotton Production and Use: Projections for 1985 and 1990. USDA, Foreign Agricultural Economic Report No. 154 (Washington, D.C., USDA).

Fischel, W. A. 1982. "The Urbanization of Agricultural Land: A Review of the National Agricultural Land Study," Land Economics, vol. 58, no. 2, pp. 236-259.

Journal of Soil and Water Conservation. 1981. vol. 36, no. 2 (March-April).

Shulstad, R. M., R. D. May, and B. E. Herrington. 1979. "Cropland Conversion Study for the Mississippi Delta Region" (Fayetteville, University of Arkansas).

_____, and J. M. Erstine. 1980. "Expansion Potential for Irrigation Within the Mississippi Delta Region" (Fayetteville, Department of Agricultural Economics and Rural Sociology, University of Arkansas).

USDA. Various Years. Agricultural Statistics (Washington, D.C., GPO).

____. 1980. Crop Production 1979 Annual Summary CrPr 2-1 (80) (Washington, D.C., January 15).

____. 1980. Basic Statistics: 1977 National Resources Inventory, Revised (Washington, D.C., Soil Conservation Service).

____. 1980. Crop Reporting Board, Acreage, CrPr 2-2 (Washington, D.C., USDA, June 27).

____. 1981. Agricultural Outlook, AO-65 (Washington, D.C., Economics and Statistics Service, May).

____. 1982. Economic Indicators of the Farm Sector: Production and Efficiency Statistics, 1980 Economic Research Service, Statistical Bulletin 679 (Washington, D.C., USDA, January).

Chapter 6

PROJECTIONS OF FERTILIZERS AND
PESTICIDES

Fertilizers

An implication of the shift by farmers toward the land-using end of
the spectrum of technologies is that fertilizer applications per acre
will grow more slowly in the future than in the past. Indeed, as table
3-1 shows, per acre applications of fertilizer grew much more slowly
after 1972 than before as farmers moved toward more land-using technologies after that date. As noted in chapter 4, the slower increase in per
acre applications of fertilizer after 1972 was consistent with the post-
1972 slowdown in yield growth. Correspondingly, our expectation that the
trend of yields to 2010 will more nearly resemble the post- than the pre-
1972 experience is based in good part on our belief that per acre applications of fertilizer will grow more slowly.

Tables 6-1 and 6-2 show, respectively, amounts of fertilizer applied
per fertilized acre and per harvested acre for corn, wheat, soybeans, and
cotton for selected years, with projections to 1985, 1990, and 2010. The
differences between the two tables reflect the percentages of harvested
acres which are fertilized.

The key elements in the projections are the percentages of land
fertilized and the amounts applied per fertilized acre. For corn and
soybeans in 1985 and 1990 we have used projections of these elements made
by Douglas.[1] We modified Douglas' projections for wheat and cotton on

---

[1]Dr. John Douglas, assistant to the manager of the Tennessee Valley
Authority. Douglas's projections are in the paper listed in the references. In conversation with one of the authors in the summer of 1980,
Douglas indicated he still was satisfied with his projections. (They
had been made in 1978.)

Table 6-1. Applications of Fertilizer Per Acre of Fertilized Land in
Corn, Wheat, Soybean and Cotton

(pounds)

| | Rate per receiving acre | | | | | | | | | | | |
| | Corn | | | Wheat | | | Soybeans | | | Cotton | | |
Year	N	P	K	N	P	K	N	P	K	N	P	K
1965	73	47	42	31	29	13	10	28	34	77	52	34
1970	112	71	72	39	30	36	14	37	51	75	55	57
1972	115	66	69	46	37	38	14	42	51	75	55	61
1973	114	64	71	48	38	36	14	42	55	73	53	62
1974	103	62	73	46	38	37	15	41	55	78	53	55
1975	105	58	67	46	35	35	15	40	53	78	50	55
1976	127	67	78	551	37	37	14	42	60	81	52	56
1977	128	68	82	53	39	41	16	45	60	78	53	52
1978	126	68	80	52	35	34	17	45	62	76	54	54
1979	135	69	84	54	38	43	16	46	67	71	50	44
1980	130	66	86	58	39	40	17	46	70	72	46	46
1985	136	70	86	60	38	42	17	50	73	67	55	50
1990	140	70	87	65	38	43	18	53	80	65	55	50
2010	145	70	90	65	38	43	20	55	85	60	55	50

Note: N = nitrogen; P = phosphorus; K = potash

Sources: 1965-79 from U.S. Department of Agriculture, Cropping Prac-
tices: Corn, Cotton, Soybeans, Wheat, 1964-70, SRS-17 (Washington, D.C.
Statistical Reporting Service, 1971); 1977 Fertilizer Situation, FS-5
(Washington, D.C., Economic Research Service, January, 1977); and 1981
Fertilizer Situation, FS-10 (Washington, D.C., Economics, Statistics and
Cooperative Service, December, 1980). Corn and soybeans in 1985 and 1990
from John Douglas, remarks presented at the Third World Fertilizer Conference, San Francisco, 1978. All other projections as described in the
text.

the basis of our judgment about recent trends. All of the projections
for 2010 are by us.

Table 6-2. Applications of Fertilizer Per Acre of Harvested Land in
Corn, Wheat, Soybeans, and Cotton

(pounds)

Year	Rate per receiving acre											
	Corn			Wheat			Soybeans			Cotton		
	N	P	K	N	P	K	N	P	K	N	p	K
1965	64	38	31	15	11	2	a	3	3	60	30	14
1970	105	64	61	24	13	7	3	10	14	54	26	21
1972	110	59	59	29	16	6	3	12	16	58	30	25
1973	106	55	57	30	17	6	3	13	18	54	29	24
1974	97	54	61	30	17	7	3	11	15	62	31	25
1975	99	50	55	29	15	7	3	10	15	51	22	18
1976	123	60	66	36	19	8	3	12	14	61	28	21
1977	123	60	67	34	17	8	4	15	18	61	27	16
1978	120	59	65	32	13	5	4	16	20	52	24	17
1979	130	61	69	35	17	8	4	17	26	50	24	12
1980	125	57	70	39	17	7	4	16	25	51	22	14
1985	133	63	73	39	19	10	4	18	26	47	28	15
1990	138	63	74	46	20	10	5	20	31	42	25	14
2010	142	63	77	49	22	10	5	22	35	36	22	13

Sources: 1965-79 from U.S. Department of Agriculture, Cropping Practices: Corn, Cotton, Soybeans, Wheat, 1964-70, SRS-17 (Washington, D.C.,
Statistical Reporting Service, 1971); 1977 Fertilizer Situation (Washington, D.C., Economic Research Service, January 1977); 1981 Fertilizer
Situation (Washington, D.C., Economics, Statistics and Cooperatives Service, December 1980). Corn and soybeans in 1985 and 1990 from John Douglas, remarks presented at the Third World Fertilizer Conference, San Francisco, 1978. All other projections as described in text.

[a]Less than one pound.

Corn

Since about 1970, nitrogen fertilizer has been applied to 94 to 97
percent of corn acreage, phosphorous to 85 to 90 percent and potash to 80
to 85 percent. Our projections for 1985, 1990, and 2010 are 98 percent
for nitrogen, 90 percent for phosphorous and 95 percent for potash.

Douglas projects very little increase in amounts of fertilizer applied per acre of corn fertilized, arguing that much, if not most, land in corn already is being fertilized at economically optimum rates. Heady (1982) expresses a similar view. If real fertilizer prices, particularly for nitrogen, rise as we expect, the economic optima will not likely change in the direction of heavier per acre applications of fertilizer. On the contrary, price movements likely will induce farmers to find ways to reduce fertilizer use without commensurate sacrifice of yield. Evidently, there already are ways of doing this which would be economically attractive with the higher fertilizer prices now in prospect.[2] Stanford (1978) believes that current yield levels can be maintained with "...an appreciable increase in efficiency" of nitrogen fertilization. He argues that amounts of nitrogen applied often are higher than necessary because farmers lack knowledge of the amount of soil organic nitrogen that will be made available to the plant by microbial action in the soil. Stanford is optimistic that current research will develop ways of predicting mineralization rates, thus enabling the farmer to incorporate this source of nitrogen into his plans and reduce the amount of purchased nitrogen applied.

Stanford also sees potential for greater efficiency by splitting applications of nitrogen in two parts: one application at planting to act as a "starter" to promote early plant growth, and a second "sidedressed" application during the growing season to ensure adequate nutrients to harvest. The alternative is to apply all the nitrogen before or at the time of planting. The advantage of split applications is that it supplies nutrients at times more in accord with the plant's need for them over the growing season, thus reducing nutrient losses that otherwise result from soil processes working on the nutrient material. However, sidedressing involves some risk. Wet fields can delay the operation, possibily causing root damage from nitrogen deficiency, or allowing the crop to grow too high to drive the application equipment through (Aldrich, 1980). Moreover, sidedressing is more difficult and more time-consuming than preplant broadcasting. When anhydrous ammonia is knifed-in between the rows,

---

[2] This account of alternatives permitting more sparing use of fertilizers is from Hemphill (1980). See also Aldrich (1980, especially pp. 263-270).

special care must be taken not to prune corn roots. The time factor is particularly relevant because the marginal value of the farmer's time typically is higher in the spring and summer than in the fall after harvest.

Split applications of nitrogen would have the highest payoff where leaching losses of nitrogen are greatest, such as in the sandy soils of the Southeast. In the Midwest, where nitrogen fertilizer use is greatest, split applications would be less attractive because soils in that area have high clay content and leaching losses are relatively low. However, if real prices of nitrogen fertilizer rise as we expect, split applications should become more attractive everywhere.

Development of new fertilizer materials which reduce nutrient losses also offers a way of achieving more efficient use of fertilizers. New forms of nitrogen in particular hold considerable promise because at present only about 50 to 70 percent of nitrogen applied is taken up by the crop (Fertilizer Institute, 1976, pp. 36-37). The high losses of N result from nitrification, the conversion of nitrogen fertilizer by soil bacteria to nitrate. Nitrate is soluble, hence susceptible to movement beyond the reach of crop roots by water. Nitrification also results in losses of N in gaseous form. Research is under way which aims at developing nitrogen fertilizer materials which retard nitrification, thus giving the crop more opportunity to use the nutrient before it is washed out of reach or volatilized. There are three approaches to slowing nitrification which presently seem to have most promise: (1) development of materials of low solubility which release nitrogen into the soil solution relatively slowly; (2) provision of a temporary barrier between the nitrogen fertilizer and the soil by coating the nitrogen with slowly decaying materials such as plastic or sulfur; (3) applications of nitrification inhibitors, chemicals which kill enough of the soil bacteria responsible for nitrification to slow the process. To date these three approaches have not proved economical for general application. However, continued research, combined with further increases in real prices of nitrogen fertilizers, should make them increasingly attractive, particularly in conditions where leaching and volatilization losses of nitrogen are high.

The spread of so-called organic farming also is a likely response to rising prices of nitrogen fertilizer. Organic farmers substitute animal manure and crop rotations which include a legume for inorganic fertilizer to supply crop requirements for nitrogen. Since corn receives far more nitrogen fertilizer than any other crop in the United States (about 40 percent of total applied N), the significance of organic farming with respect to use of nitrogen fertilizer depends largely on its potential use in corn production. Many farmers in the United States already practice organic farming, and according to a USDA report (July 1980) the technology has promise for more widespread use, particularly if real prices of nitrogen fertilizer rise. There is nothing in the report, however, to suggest that organic farming is likely to substitute in a major way for current fertilizer management practices. The report examined the economics of organic and conventional farming in the Midwest using crop budget data developed by the USDA and Oklahoma State University. Three organic farming systems were examined: a four-year alfalfa-corn-soybean-oats rotation, a five-year alfalfa-alfalfa-corn-soybean-oats rotation, and a seven-year alfalfa-alfalfa-corn-soybean-corn-soybean-alfalfa rotation. The conventional system was a two-year corn-soybean rotation. The conventional system produced a larger annual return above variable costs than any of the organic systems. A main reason for this was that with the organic farming systems only about 57 percent of total crop acreage on average was in corn and soybeans, while the conventional system had 100 percent of its acreage in these relatively high valued crops.[3]

The USDA report concludes that the spread of organic farming is limited by "...the lack of an adequate and economical supply of organic wastes and residues and/or because soil nutrients and climatic conditions are not suitable for successful and profitable organic farming. Based on our observations, the greatest opportunity for organic farming will probably be on small farms and on larger mixed crop/livestock farms with large numbers of animal units" (USDA, July 1980, pp. 46-47).

---

[3]The study assumed that yields were 10 percent lower with the organic systems. Even if no difference in yields were assumed, however, returns above variable costs still would have been higher with the conventional system.

Lockeretz, Shearer, and Kohl (1981) compared organic and conventional farms in the Corn Belt over the five years 1974-1978. They found that conventional farms generally had higher yields and gross revenues. However, variable costs also were higher on the conventional farms so their advantage in net revenues was small—only 2 percent over the five years as a whole. This is a smaller difference that was found in the USDA study. A main reason, evidently, is that many of the conventional farms studied by Lockeretz and coauthors were mixed crop-animal enterprises. These farms typically have some land in relatively low value forage crops. As noted above, the USDA study assumed a corn-soybean rotation.

We conclude that the economics of organic farming do not favor a large scale shift to such systems. Accordingly, we view the prospects for organic farming as sufficiently good to restrain the increase in per acre application of nitrogen fertilizer, but not so good as to cause a decline.[4]

Wheat

The percentage of wheat land fertilized is substantially less than that for corn, and has shown no trend in the 1970s. Sixty to 65 percent of wheat acreage receives nitrogen, 40 to 50 percent receives phosphorous and 15 to 20 percent receives potash. Douglas projects an increase in all these percentages, particularly for nitrogen. The rationale for this is not clear, and we believe the outlook for fertilizer prices makes it doubtful. Accordingly, we have projected a more modest increase in the percentages of wheat land fertilized.

Applications of nitrogen per acre of wheat land fertilized rose in the 1970s, and we project a continuation, although at a slower rate than Douglas. Phosphorous applications, however, showed no trend and that for potash was only slightly rising. Our projections of phosphorous and potash applications per acre of wheat land receiving these nutrients reflect these patterns.

---

[4]According to the USDA (July 1980) report, organic farming may deplete soil stocks of phosphorous and potash, indicating that after a few years of organic farming, these nutrients would have to be applied on a regular basis. Adoption of organic farming, therefore, particularly on the modest scale we anticipate, would not restrain per acre applications of P and K.

Soybeans

As a legume, soybeans biologically fix most of the nitrogen they require. Consequently, the percentage of soybean acreage receiving nitrogen fertilizer and the per acre amount applied are both small. Considerably greater quantities of P and K are applied to soybean land which receives any of these materials, and application rates, particularly of K, rose in the 1970s. Douglas projects a continuation of these increases. However, the percentage of soybean land which receives P and K is small, and Douglas projects little increase in it. Consequently, the projected amounts of P and K per acre of land in soybeans are small and show little increase. Douglas argues that this pattern will characterize fertilizer use on soybeans absent a significant breakthrough in the agronomics of soybean production.

Cotton

The percentages of cotton acres receiving N, P and K declined slightly in the 1970s, and the amounts applied per receiving acre were stable to slightly declining. This behavior reflected the shift of cotton acreage to the Southern Plains--Texas and Oklahoma--where percentages of acres receiving and amounts per receiving acre are low, from the Southeast and Mississippi Delta, where percentages and amounts are high. Our projections of the regional distribution of cotton production and acreage assume that the shift to the Southern Plains will continue. Our projections of fertilizer use are generally consistent with this shift.

Other Uses of Fertilizer

In addition to its use on corn, wheat, soybeans and cotton, fertilizer also is applied to other field crops, hay and pasture, fruits and vegetables, forests, home gardens and lawns, golf courses, roadways and for other miscellaneous uses (Douglas, 1978). In the second half of the 1970s, these other uses took about 43 percent of all the nitrogen applied, 39 percent of the phosphorous and about 42 percent of the potash. Douglas's projections indicate that the percentage of nitrogen taken by these other uses would rise to 50 percent in 1985 and to 52 percent in 1990. For P and K, Douglas projects only marginal increases in the percentages taken by these other uses.

We think it likely that the price elasticity of demand for these
other uses of fertilizer is lower than it is for corn, wheat, soybeans and
cotton. The average homeowner, for example, is likely to be less sensitive to the price of fertilizer in deciding how much to apply to his garden or lawn than the farmer in deciding how much to apply to his corn crop.
Douglas's projections for the four main crops are consistent with this line
of reasoning. We project the amounts of N, P and K taken by other uses by
assuming that they will be in the same relation to our projected amounts
of each used on the main crops as in Douglas's projections. Table 6-3
shows the projections for each crop and for all other uses.

Pesticides

Data Problems

Projections of pesticide use in the United States are on an uncertain
footing because of wide and unexplained discrepancies among the data showing present amounts used and trends in amounts. Perhaps the most widely
cited data are from surveys of pesticide use conducted by the USDA in 1964,
1966, 1971 and 1976. These data, however, are at variance with those collected by others. This is indicated in table 6-4. Von Rumker and coauthors carefully examined the differences between their estimates and
those of the USDA. They concluded that some of the differences could be
accounted for by increased use from 1971 to 1972, e.g., methyl parathion
and toxaphene on cotton and soybeans; also alachlor. The greater part of
the differences, however, could not be explained in this way. Von Rumker
and coauthors note that the USDA estimates are from a survey using a
"...mammoth form consisting of 56 pages and 487 questions..." (p. 27) which
required approximately 3 hours to complete. Twenty-eight of the 487 master
questions had to do with pesticides. According to von Rumker and coauthors,
even a person highly trained in the field of pesticides, with knowledge of
the many different kinds and formulations of materials used, would have
difficulty completing the form accurately.

The estimates of the von Rumker study are based on comprehensive surveys of agricultural extension specialists in the various states, Directors
of EPA Community Pesticides Studies Projects in a number of states, and

Table 6-3. Amounts of Fertilizer Applied to Land in Corn, Wheat,
Soybeans, Cotton and for All Other Uses
(millions of metric tons)

Crop	1977/79			2010		
	N	P	K	N	P	K
Corn	3.97	1.93	2.15	5.96	2.64	3.24
Wheat	.87	.41	.18	2.01	.90	.41
Soybeans	.10	.30	.61	.25	1.09	1.73
Cotton	.31	.16	.08	.25	.15	.09
Subtotal	5.25	2.80	3.02	8.47	4.78	5.47
All Other	4.18	2.11	2.28	9.18	3.46	4.48
Total	9.43	4.91	5.30	17.65	8.24	9.95

Source: 1977-79 from U.S. Department of Agriculture, 1981 Fertilizer
Situation, FS-10 (Washington, D.C., Economics, Statistics and Cooperatives
Service, December 1980). Projections as described in text.

pesticide manufacturers. For 25 most important pesticides which they studied intensively, von Rumker and coauthors concluded that their estimates
were accurate within ± 10 percent for quantities over 10 million pounds
and generally within ± 1 million pounds for quantities less than 10 million
pounds.

Von Rumker and coauthors conclude that their estimates are more accurate than those of the USDA, and we accept this judgment. However, the
USDA estimates have the major advantage of covering several years (1964,
1966, 1971 and 1976), thus providing information about changes over time.
This information, of course, is particularly valuable in making judgments
about trends in future use of pesticides.

We resolved the dilemma posed by differences in the pesticide use
data by rejecting projections of specific quantities of various kinds of
pesticides. Instead, we project directions of change, and marshal evidence for judgments of whether the changes are likely to be much or little.
For this purpose we believe the USDA data are adequate and useful, especially since they were collected on a consistent basis, even though the

Table 6-4. Contrasting Estimates of Amounts of Herbicides and
Insecticides Used in the United States
(millions of pounds active ingredients)

	(1) von Rumker and coauthors (1972)	(2) USDA (1971)	(3) Col. (1) – Col. (2)
Insecticides			
Carbaryl	19.0	11.2	1.70
Carbofuran	5.0	2.8	1.79
Disulfoton	4.9	2.8	1.75
Methyl parathion	39.7	27.1	1.46
Parathion	10.0	7.0	1.43
Toxaphene	57.0	31.9	1.79
Total	135.6	82.8	1.64
Herbicides			
Alachlor	21.0	14.0	1.50
Atrazine	72.0	53.9	1.34
2,4-D	36.0	30.5	1.18
Trifluralin	16.8	10.3	1.63
Total	145.8	108.7	1.34
Insecticides on			
Cotton	105.0 (approx.)	73.4	1.43

Sources: R. von Rumker, E.W. Lawless, and A.F. Meines, with K.A.
Lawrence, G.L. Kelso, and F. Horay, Production, Distribution, Use and
Environmental Impact Potential of Selected Pesticides (Washington, D.C.,
EPA, 1975); USDA from P. Andrilenas, Farmers' Use of Pesticides in 1971 –
Quantities, Agricultural Economics Report no. 252, ERS-USDA (Washington,
D.C., USDA, 1974).

totals for each year may be questionable. We focus only on insecticides
and herbicides used in crop production.[5]

---

[5]Fungicides and other pesticides accounted for 14 percent of the
quantity of all pesticides used on crops in 1976, down from 19 percent
in 1971. The share of these materials among all pesticides used in crops
of interest to this study is much less than these percentages indicate.
Moreover, these materials in general pose fewer threats to the environment than insecticides and herbicides.

Insecticides

Tables 6-5 through 6-9 show data on insecticides used on crops. For our purposes the most important information in the tables is the following:

1. The total amount of insecticides applied increased 5.4 percent from 1971 to 1976, much less than the percentage increase in quantity of herbicides.

2. Insecticides applied to cotton accounted for 48 percent of total crop use of insecticides in 1971 and for 40 percent in 1976.

3. Corn accounted for 17 percent of total crop use of insecticides in 1971 and for 20 percent in 1976.

4. In 1976, 83 percent of the insecticides applied to cotton was used in the Mississippi Delta (50.9 percent) and the Southeast (32.1 percent). Together, these two regions accounted for one-third of all the insecticides used on crops in the entire nation.

5. The Corn Belt, Northern Plains and Lake States accounted, respectively, for 44.1 percent, 25.5 percent and 15.6 percent of all insecticides applied to corn in 1976.

6. Amounts of insecticides applied per acre of land in cotton and corn declined from 1971 to 1976.

7. Organochlorine compounds (e.g. DDT, toxaphene) declined from 46 percent of all insecticides applied to crops in 1971 to 29 percent in 1976. Organophosphorus compounds (e.g. methyl parathion) increased from 40 to 49 percent and carbamates (e.g. carbaryl, carbofuran) increased from 14 percent to 19 percent.[6]

These statements suggest that unless there is reason to believe that insecticides will be applied to sharply higher percentages of land in wheat and soybeans, the direction of total insecticide use on crops in the future will depend overwhelmingly on trends in use on cotton in the Southeast and Delta and on corn in the three principal corn producing regions. Further, if the substitution of organophosphorous for organochlorine compounds continues, problems resulting from persistence of insecticides in the environment will diminish in relative importance and those of acute toxic effects on humans and animals will increase.

─────────────────

[6]These data are not in tables 6-5 through 6-9. They are from Eichers, Andrilenas, and Anderson (1978, p. 16).

Table 6-5. Pesticide Use on Crops in 1971 and 1976

	Amounts applied (millions of pounds active ingredients)		Acres treated (millions)	
	1971	1976	1971	1976
Herbicides	224.0	394.3	157.8	196.6
Insecticides	153.8	162.1	56.7	74.9
Fungicides	39.6	43.2	8.5	10.5
Other	46.3	50.2	10.0	11.6
Total	463.7	649.8	n.a.	n.a.

Note: n.a. = not available.

Source: T.R. Eichers, P. Andrilenas, and T.W. Anderson, Farmers' Use of Pesticides in 1976, USDA-ESCS, Agricultural Economic Report no. 418 (Washington, D.C., USDA, 1978), p. 6.

Table 6-6. Percentages of Acres on Which Pesticides Were Used in 1976, Major Crops

	Herbicides	Insecticides	Fungicides	Other pesticides	Any pesticides
Corn	90	38	1	1	92
Cotton	84	60	9	34	95
Wheat	38	14	1	a	48
Sorghum	51	27	–	a	56
Other grains[b]	35	5	2	–	41
Soybeans	88	7	3	1	90

Source: T.R. Eichers, P. Andrilenas, and T.W. Anderson, Farmers' Use of Pesticides in 1976, USDA-ESCS, Agricultural Economic Report no. 418 (Washington, D.C., USDA, 1978), p. 7.

[a]Less than .5 percent.

[b]Oats, rye, and barley

Total 6-7. Applications of Herbicides and Insecticides to Main Crops in 1971 and 1976 (active ingredients)

Crop	Herbicides 1971 Total (mill. lb)	Herbicides 1971 Per acre treated (lb)	Herbicides 1976 Total (mill. lb)	Herbicides 1976 Per acre treated (lb)	Insecticides 1971 Total (mill. lb)	Insecticides 1971 Per acre treated (lb)	Insecticides 1976 Total (mill. lb)	Insecticides 1976 Per acre treated (lb)
Corn	101.1	1.7	207.1	2.7	25.5	1.2	32.0	1.0
Cotton	19.6	1.9	18.3	1.9	73.4	9.8	64.1	9.2
Wheat	11.6	.5	21.9	.7	1.7	.4	7.2	.6
Sorghum	11.5	1.2	15.7	1.7	5.7	.7	4.6	.9
Other grains[a]	5.4	.5	5.5	.5	.8	.7	1.8	1.2
Soybeans	36.5	1.2	81.1	1.8	5.6	1.6	7.9	2.3
Total of these crops	185.7	n.a.	349.6	n.a.	112.7	n.a.	117.6	n.a.
Total all crops	224.0	1.4	394.3	2.0	153.8	2.7	162.1	2.2

Note: n.a. = not available

Source: T.R. Eichers, P. Andrilenas, and T.W. Anderson, Farmers' Use of Pesticides in 1976, USDA-ESCS, Agricultural Economic Report no. 418 (Washington, D.C., USDA), pp. 9 and 15.

[a]Oats, rye, and barley.

Table 6-8. Types and Amounts of Insecticides Applied to Main Crops

Crop	Million lb	Percentage of total
Corn		
Carbofuran	9.9	30.9
Phorate	5.8	18.1
Fonofos	5.0	15.6
Other	11.3	35.4
Total	32.0	100.0
Cotton		
Toxaphene	26.3	41.0
Methyl parathion	20.0	31.2
Other	17.8	27.8
Total	64.1	100.0
Wheat		
Parathion	3.1	43.1
Disulfoton	1.8	25.0
Methyl parathion	1.2	16.7
Other	1.1	15.2
Total	7.2	100.0
Soybeans		
Carbaryl	3.7	46.8
Toxaphene	2.2	27.9
Other	2.0	25.3
Total	7.9	100.0
Sorghum		
Parathion	1.2	26.1
Disulfoton	1.1	23.9
Toxaphene	1.0	21.7
Other	1.3	28.3
Total	4.6	100.0

Source: T.R. Eichers, P.A. Andrilenas, and T.W. Anderson, Farmers' Use of Pesticides in 1976, USDA-ESCS Agricultural Economics Report no. 418 (Washington, D.C., USDA, 1978), p. 18.

Table 6-9. Herbicides and Insecticides Applied to Main Crops, by Region in 1976 (million lb)

Region	Corn Herbicides	Corn Insecticides	Cotton Herbicides	Cotton Insecticides	Wheat Herbicides	Wheat Insecticides	Soybeans Herbicides	Soybeans Insecticides	Sorghum Herbicides	Sorghum Insecticides
Northeast	10.93	1.02	--	--	.01	.02	1.32	.35	.13	--
Appalachia	19.09	.94	.75	4.09	.08	.17	8.21	.87	1.50	--
Southeast	8.13	.97	1.04	20.58	--	--	6.37	6.18	.05	.10
Delta	.39	.02	11.56	32.65	.06	--	15.24	.17	.42	.49
Lake states	33.91	5.00	--	--	2.41	.01	6.05	.02	.02	--
Corn Belt	108.04	14.09	--	--	.06	.48	41.51	.12	1.30	.30
Northern Plains	22.81	8.17	2.76	2.46	6.22	.40	2.35	a	7.94	2.20
Southern Plains	1.66	1.25	1.30	3.34	.98	4.49	.01	.15	4.10	1.37
Mountain	1.19	.37	.90	1.02	3.92	.41	--	--	.22	.06
Pacific	.91	.14			8.15	1.27	--	--	.05	.09
United States	207.06	31.98	18.31	64.14	21.88	7.24	81.06	7.87	15.72	4.60

Source: T.R. Eichers, P.A. Andrilenas, and T.W. Anderson, Farmers' Use of Pesticides in 1976, USDA-ESCS Agricultural Economics Report no. 418 (Washington, D.C., USDA, 1978), pp. 13 and 19.

^aLess than 5,000 pounds.

83

Table 6-10. Types and Amounts of Herbicides Applied to Main Crops
in 1976

Crop	Millions lb	Percent of total
Corn		
Atrazine	83.8	40.5
Alachlor	58.2	28.1
Butylute	24.3	11.7
Other	40.8	19.7
Total	207.1	100.0
Cotton		
Trifluralin	7.0	38.3
Flurometuron	5.3	29.0
Other	6.0	32.7
Total	18.3	100.0
Wheat		
2, 4-D	15.5	70.8
Other	6.4	29.2
Total	21.9	100.0
Soybeans		
Alachor	29.6	36.5
Trifluralin	21.1	26.0
Other	30.4	37.5
Total	81.1	100.0
Sorghum		
Atrazine	6.5	41.4
Propazine	3.9	24.8
Propachlor	3.1	19.8
Other	2.2	14.0
Total	15.7	100.0
Other grain[a]		
2, 4-D	3.8	69.1
Other	1.7	30.9
Total	5.5	100.0

Source: T.R. Eichers, P.A. Andrilenas, and T.W. Anderson, Farmers'
Use of Pesticides in 1976, USDA-ESCS Agricultural Economics Report no.
418 (Washington, D.C., USDA, 1978), p. 12.

[a]Oats, rye, and barley.

Wheat. The percentage of wheat acres treated with insecticides and the amount applied per treated acre are small (tables 6-6 and 6-7) because wheat generally is not attacked by insects on a scale requiring a response. The principal insect pests of wheat are the Hessian fly, greenbug, wheat stem sawfly, armyworms and cutworms (Office of Technology Assessment, 1980). According to the OTA, these pests present "serious to occasional" threats to wheat. However, the damages typically are sufficiently small that it is not economical for farmers to invest heavily in insecticide treatments. By 1971, 95 percent of the insecticides used on wheat were organophosphorous compounds (OTA, 1980, p. 21). Use of cultivars with resistance to insects is a major control practice in wheat production. According to the OTA (1980, p. 27), it is in fact the most effective means for controlling the Hessian fly and wheat stem sawfly.

In judging the development of control strategies for wheat insects into the 1990s, the OTA concludes that the role of insecticides likely will diminish. It is expected that more accurate determination of economic thresholds will reduce unneeded applications,[7] and improvements in application equipment will result in desired levels of control with smaller amounts of insecticides.

The OTA report thus gives no reason to believe that the percentage of wheat acres treated with insecticides or the amounts applied per treated acre are likely to increase. On the contrary, the thrust of the argument in the report is that insecticide use on wheat is likely to decline. In this connection, we think it significant that in a major study by the National Academy of Sciences of pesticide practices in the United States, separate volumes were prepared dealing with cotton and with corn/soybeans, but none with wheat. The EPA commissioned a study of alternatives for reducing insecticides applied to cotton and corn (Pimental and coauthors, 1977), but evidently has seen no need for a similar study for wheat.

Soybeans. The low percentage of soybean land treated with insecticides (table 6-6) masks a wide difference between the percentages for soybean land in the Corn Belt and Delta on the one hand and in the Southeast

---

[7]The economic threshold is reached at the point where the amount the farmer would save by applying insecticides to reduce insect damage is just equal to the cost of the insecticides.

on the other. According to the OTA (1980) report, 1 to 10 percent of soybean land in the Corn Belt is treated annually for insects. In Arkansas,
the figure in 1977 was 6 percent and in Louisiana and Mississippi 90 percent and 75 percent respectively. However, the report notes that the figures for Louisiana and Mississippi "were very high compared with normal
usage." In the Southeast the percentage of soybean land treated was 70-75
percent, which evidently is typical for that region.

In 1977-79 the Southeast had 8.5 percent of the nation's land in soybeans. Our projections of production and land use imply that the Southeast's share of soybean land would rise to about 13 percent by 2010. The
Delta's share would rise slightly and that of the Corn Belt would decline.
These regional shifts in land in soybeans would imply a small increase
in the nationwide percentage of soybean land receiving insecticides, assuming persistence of the regional differences in percentages noted above.

According to the OTA (1980) report, there currently are only 2 major
insect pests of soybeans in the Corn Belt, the Mexican bean beetle and the
green cloverworm. Each now presents a "moderate" threat. By the 1990s,
the OTA report judges that the Mexican bean beetle may constitute a "high"
threat, but that the threat from the green cloverworm would remain moderate. The threats from all other insect pests of soybeans in the Corn Belt
are judged now to be low to very low, and to remain that way into the 1990s.
The OTA report has little to say concerning future practices for control of
soybean insect pests in the Corn Belt. It notes, however, that insect-
tolerant varieties of soybeans have been identified and strong effort is
underway to introduce this characteristic into commercial varieties of
soybeans. According to the report, total resistance is not necessary for
soybeans in the Corn Belt. Varieties with 50 to 75 percent tolerance to
leaf-feeding caterpillars and beetles would have little need for insecticides.

The high percentage of soybean land in the Southeast which receives
insecticides indicates that the insect problem in that region is more severe than in the Corn Belt or Delta. The OTA report concludes that insecticides will continue to play a role in control of insect pests of
soybeans in the Southeast and in the Delta. The report notes, however,
that soybean varieties with resistance to insects of those regions already

have been developed and asserts that a much greater commitment of resources to research along this line would have high payoff over the long term. The report also sees promise in the present development of biological techniques for control of soybean insects in the Southeast and Delta, and urges that further development of these techniques should receive high priority.

In summary, although the small relative shift of soybean production to the Southeast would tend to increase slightly the nationwide percentage of soybean land receiving insecticides, the development of alternative control practices in the Southeast, Delta, and Corn Belt looks promising enough to offset this tendency. Unless there is an outbreak of some insect pest of soybeans not presently identified as a serious threat, it appears unlikely that the percentage of soybean acreage receiving insecticides or the amounts applied per receiving acre will increase much, if at all.

Corn. The principal insect pests of corn are the northern and western rootworm, European cornborers and the black cutworm (OTA, 1980). Of these, the corn rootworm is easily the most important, and most of the insecticides applied to corn are for control of this insect. It is not too much to say that the future trend in use of insecticides on corn will depend primarily on development of practices which modify or substitute for insecticides currently used to control the rootworm.

At present the principal alternative to insecticides for control of the corn rootworm is rotation with another crop, primarily soybeans, since the rootworm must have corn roots to survive. In deciding whether to adopt this practice the farmer will consider and balance the economics of the corn-soybean rotation against its advantages for control of the rootworm. According to a study by Miranowski (1979), the corn-soybean rotation, with no use of insecticides, was in fact the most economical means of controlling the corn rootworm in the Corn Belt under cost-price relationships prevailing the late 1970s.

Part of the advantage of control by rotation is that it slows the buildup of genetic resistance of the rootworm to insecticides, thus extending the useful economic life of currently used chemicals. This advantage likely will increase in the future because of increasing costs of developing new insecticides (OTA, 1980, Part 3, p. 43).

So-called insect monitoring programs, or "scouting" are coming into increasing use for control of insect pests of corn. Scouting is the key component of so-called integrated pest management (IPM) in corn. The principle of IPM as currently practiced is to apply insecticides only when, where and in the amount needed to prevent pest damage from crossing the economic threshold. The programs rely on "scouts" to provide information about the number of insects in the field and about the threat they present at various stages of their life cycle and that of the plant. Typically, these programs result in more sparing use of insecticides for a given level of insect control than practices based on rules-of-thumb about how much insecticide to apply. In effect, IPM substitutes knowledge for chemicals at the margin to control insects.

Limited or experimental scouting programs have been mounted by the Agricultural Extension Service in the Corn Belt and have been well received by farmers. The spread of these programs could significantly reduce the quantities of insecticides needed for effective control of the corn rootworm (OTA, 1980, Part 3, pp. 37-38).[8]

The development of cultivars of corn with greater insect resistance than those now used also has good potential as an insect control strategy. The OTA expects that this strategy in fact will pay off by the 1990s, resulting in a reduction from 1976 levels in the amount of insecticides applied to corn (OTA, 1980, Part 3, p. 51).

We expect a considerable expansion of corn land in conservation tillage, for reasons detailed below in the discussion of herbicides. According to the OTA report this could increase the threat to corn of insects now only of minor importance. Conservation tillage leaves residue from the previous crop on the soil surface, creating a hospital environment for insects, particularly the black cutworm and armyworm, which could not

---

[8]David Pimentel (personal communication) is less sanguine than the OTA about the potential for scouting to reduce insecticide use on corn. He points out that only about one pound of active ingredient is used per acre, indicating that the cost of treatment is low. Hence the saving from adoption of IPM also is low. John Miranowski (1979) takes a similar view of the potential for IPM in corn, arguing that the rate of adoption of the technique likely will be slow unless insecticide prices rise substantially relative to costs of IPM.

flourish in a clean tilled field. To the extent that this occurs, it would offset the effects of increased corn-soybean rotation and scouting in reducing the amount of insecticides applied to corn. It is worth noting in this connection, however, that the amount of land in conservation tillage doubled between 1972 and 1976 (No-Till Farmer, March 1981) but total insecticide use increased less than 6 percent (table 6-5). If conservation tillage require significantly more insecticides than conventional tillage, one would have expected insecticide use to increase more. Conservation tillage definitely requires more herbicides than conventional tillage, and harbicide use rose sharply from 1971 to 1976 (table 6-5).

Our projection of land in corn indicates a 30 percent increase from 71.3 million acres in 1976 to 93 million acres in 2010. We expect a decline in the amount of insecticides applied per receiving acre of corn, or in the percentage of land receiving insecticides, or in some combination of the two, to offset a considerable part, if not all, of the increase in land in corn.[9] Accordingly, we expect little if any increase in the total amount of insecticides applied to corn over the next several decades.

Cotton. We expect a significant decline from 1976 to 2010 in the total amount of insecticides applied to cotton. There are two reasons: (1) a marked shift in the regional distribution of land in cotton from the Mississippi Delta, where rates per receiving acre are high, to Texas where rates per receiving acre are low; (2) a decline in present high rates per receiving acre in the Delta and in the Southeast.

Because the climate in the High Plains of Texas is hostile to cotton insect pests, cotton farmers in that region can produce a crop in most years without applying any insecticides (National Academy of Sciences, 1975). Since 40 to 45 percent of the cotton acreage in Texas is in the High Plains, the low per acre amounts of insecticides applied in that area keep the average amount for the state as a whole low.

But application rates elsewhere in Texas also are low in comparison with rates in the Southeast and Delta. The reasons for this are complex, and what follows does not pretend to be a complete account. There appear

---

[9]According to table 6-7, a decline in the amount of insecticides per receiving acre of corn already had occurred from 1971 to 1976.

to be two key components to the explanation, however. One is that, beginning in the 1950s, the key insect pests of cotton in Texas, particularly
the boll weevil, began to develop genetic resistance to organochlorine
insecticides.[10] Subsequently, the cotton bollworm and tobacco budworm
also developed resistance to these materials. Farmers switched to organophosphorous compounds, at first with good success, but subsequently the
tobacco budworm developed resistance to these as well. The result was
that when farmers sprayed with organophosphates to control the boll weevil
or other primary insect pests, they also killed off predators of the budworm, which then was able to, and often did, wreak considerable damage.

Researchers went to work on this problem and came up with insect
management practices which usually give adequate control with relatively
little reliance on insecticides. These practices are built around the
fact that, in most instances, the tobacco budworm is a secondary pest of
cotton whose numbers can be kept below economically damaging amounts by
reliance on insect predators of the budworm. The key is not to destroy
the predators. This is avoided by spraying either very early in the cotton
growing season with one or two treatments to kill boll weevils surviving
the winter or for flea-hopper infestation, or very late in the season to
get weevils preparing to overwinter. These approaches avoid outbreaks
of the budworm-bollworm complex so well that multiapplications of insecticides no longer are needed or considered good practice by most cotton producers in Texas.

This mode of attack has benefited also from the development and adoption of "short-season" varieties of cotton in Texas. These varieties mature earlier than the varieties they have replaced, getting them through
the stage of development when they are most vulnerable to insect attack
before insect populations swell to threatening numbers.

The other major component to the explanation of why per acre applications of insecticides to cotton are relatively low in Texas is that the
yield potential of much cotton land in the state is less than in the Delta

---

[10]This account of developing insect resistance to insecticides is
based on OTA (1980, Part 8), and on personal communications with Ray Frisbie, Michael McWhorter, Robert Metzer, Knox Walker, and John G. Thomas,
all of Texas A&M University.

and Southeast (National Academy of Sciences, 1975, p. 65). A main reason for this is inadequate water in the semiarid cotton growing region of the state.[11] Consequently, the yield increase that Texas cotton farmers would obtain if they put on as much insecticides as farmers in the Delta and Southeast would not justify the additional cost. Moreover, cottin in the Southeast and Delta encounters stronger competition than in Texas from other high value crops, e.g., soybeans. To meet the competition, farmers in those regions must get higher cotton yields than in Texas, which implies higher per acre use of insecticides.[12]

In 1976, .51 pounds of insecticides were applied per acre of cotton in Texas and Oklahoma. The figures were 10.96 pounds in the Delta and 24.92 pounds in the Southeast.[13] The average application rate for the three regions was 6.44 pounds per acre. Our projections of the regional distribution of land in cotton would reduce this average to 3.29 pounds in 2010, even if per acre applications in each region remained at the 1976 level. In this case, the amounts of insecticides used on cotton in the three regions would decline from 55.7 million pounds in 1976 to 38.8 million pounds in 2010.[14]

Per acre applications of insecticides to cotton in Texas already are so low that they probably will not go much lower. They could rise if the boll weevil or other primary pests were to develop resistance to the organophosphorous compounds now used, requiring increased applications for a given amount of control. This ultimately would be a self-defeating

---

[11]John G. Thomas, personal communication.

[12]Cotton yields in Texas averaged 345 pounds per acre in 1977/79. In the Delta they were 566 pounds and in the Southeast 482 pounds (USDA, January 15, 1980).

[13]Total insecticides applied in each region are from Eichers, Andrilenas and Anderson (1978). Land in cotton is from the USDA (1979).

[14]At the beginning of this discussion of pesticides it was noted that the USDA estimates of use differ significantly from others and are probably too low. This would not likely affect the proportionate distribution of use among regions, however. Consequently, while the above estimates of per acre insecticide use on cotton in each region may be too low, estimates of the effect of a shift to the Southern Plains on the average application rate for the three regions should be accurate enough.

response, however. And the demonstrated capacity of the research establishment and cotton producers in Texas to develop insect control strategies requiring only sparing use of insecticides suggests they would not
respond to increasing boll weevil resistance simply by increasing per
acre application rates. On this ground, we think it unlikely that rates
will rise much if at all in Texas. Even if they do increase, they are so
low relative to those in the Southeast and the Delta that the projected
shift of cotton acreage to Texas still would reduce the average application rate and total amount of insecticides used on cotton in the three
regions.

As noted above, the higher yield potential of cotton land in the
Southeast and Delta compared to Texas makes it economical for farmers in
those two regions to apply more insecticides per acre than Texas farmers
do. This suggests that per acre applications in the Southeast and Delta
will continue to be well above rates in Texas for the foreseeable future.
Nonetheless, there is reason to believe that they will decline from the
1976 rates. Indeed some observers believe they already have, at least
in some places.[15] Where amounts applied per acre have declined, one of
the reasons is the increased use of synthetic pyrethroids. This material,
effective against the budworm-bollworm complex where other insecticides
are not, is applied in much smaller amounts than the material it replaces.
However, according to Hamer, the apparent decline after 1976 in pounds of
insecticides applied to cotton in Mississippi was not just because of the
increased use of pyrethroids.

Cotton farmers in the Delta and Southeast have rapidly increased the
use of cotton pest management consultants in the last few years. According to Hamer, these consultants at the beginning of the 1980s provided pest
management services to operators of about 85 percent or 730,000 to 750,000
acres of cotton land in the Delta region of Mississippi. About 75 percent
of cotton acreage in Alabama is said to be scouted for insects by exten-

---

[15]In conversation with one of the authors in the winter of 1980,
James Hamer, Agricultural Extension specialist at Mississippi State University involved in cotton pest management programs, stated that insecticide use on cotton in Mississippi probably has declined from the 1976
level.

tion trained people and about 18 percent by professional consultants.[16]
Most farmers in South Carolina use scouts, according to one source, and in
Georgia about 85 percent of cotton acreage is scouted, according to another.[17] While data are not available showing the effects of the spread of
scouting on insecticide application rates, the presumption is that rates
required for a given amount of protection decline.

As noted above, one of the key components of cotton insect management
strategies in Texas is use of "short-season," or early maturing varieties,
the advantage of these varieties being that they pass through the stages
of greatest vulnerability to insect attack before insect populations reach
threatening size. It is natural to ask whether these varieties could not
also be used in the Delta and Southeast to reduce requirements for insecticides.

Research is under way to develop short-season varieties adapted to
those regions. Some are now used in Mississippi, but the acreage is very
small. There are some important obstacles to widespread adoption of these
varieties, however. The Texas varieties were developed for use in a dry
climate, a condition not met in the Delta and Southeast. Not only is the
climate wetter in those regions, but it also is more variable. The traditional, long-season varieties used in the Delta and Southeast are more
resilient in recovering from extreme weather stress than the short-season
varieties of Texas. Where the short-season varieties have been tried
in experiment stations in the Southeast, yields have been lower than with
the traditional varieties. This is not a serious disadvantage in Texas
where total costs per acre generally are lower than in the Southeast, but
in the latter region high yields are necessary to offset higher costs.[18]

---

[16]Communication with John French, pest management specialist with
the Cooperative Extension Service, Auburn University.

[17]Communication with Donald Johnson, Department of Entomology and
Zoology, Clemson University, and with William Lambert, Coastal Plain
Experiment Station, Tifton, Georgia.

[18]This account of obstacles to the adoption of short-season cotton
varieties in the Delta and Southeast is based on discussion with Donald
Johnson, Clemson University.

The USDA is sponsoring research on three approaches to management of cotton insect pests: boll weevil eradication (BWE), optimum pest management (OPM), and current insect control (CIC) practices. The objective is to describe relevant features of the alternative approaches and estimate their economic and environmental impacts, both at the farm level and throughout the cotton belt. The studies are expected to serve as a basis for choosing among insect management strategies and for policies to promote the chosen strategies, or mix of strategies. Since the results of the studies are not available at this writing, it is impossible to know what their implications may be with respect to trends in quantities of insecticides applied to cotton. At least one participant in the studies, however, asserts that the main difference between the OPM strategy and the CIC strategy is that OPM would increase the role of knowledge about plant-insect interactions and reduce the role of insecticides, primarily because the rising costs of the latter make it imperative to use them more sparingly.[19]

There is much other research underway, most of it pointed toward finding alternatives to insecticides for insect control, alternatives which would not likely eliminate the use of insecticides, but which would diminish their role. One approach now receiving much attention is development of insect resistant varieties. A variety with some resistance to the boll weevil, tobacco budworm, and fleahopper already has been developed in Texas.[20] Some use has been made of bacteria and viruses for control of the budworm complex in the Delta, although not in the Southeast, and research is continuing along these lines.[21]

While some of the various alternative strategies now being explored through research probably will not prove economically viable at the farm level, others almost surely will. We believe this outcome is sufficiently likely to justify the conclusion that per acre applications of insecticides to cotton in the Southeast and Delta will decline from 1976 levels, probably substantially by 2010. Since we project a reduction in total

---

[19] Donald Johnson, Clemson University, personal communication.

[20] Raymond Frisbie, Texas A&M University, personal communication.

[21] Donald Johnson, Clemson University.

cotton acreage in those regions, we expect the total insecticide load also to decline. When this is combined with the projected shift of cotton acreage to Texas, the implication is a substantial drop for the three regions combined both in per acre applications of insecticides to cotton and in the total quantity of insecticides used.

On balance, we expect total insecticide use on grains, soybeans and cotton to decline, primarily, but not exclusively, because of declining use on cotton. A study by Headley (1981) supports this conclusion with respect to grains and soybeans. (The study was not concerned with cotton.) As part of the study, Headley surveyed U.S. agricultural extension and research workers familiar with pest management practices in grains and soybeans. The thirty-nine responses indicated a consensus that chemical insecticides would continue to play a major role into the 1990s, but that the trend of use would be declining. Resistant crop varieties, already of major importance for control of grain and soybean insects, would be on a rising trend. All other biological control, e.g., deliberate use of parasites and predators, bacteria, viruses, and phermones, would remain of minor importance.

Herbicides

Importance of Conservation Tillage.[22] Table 6-5 shows a substantial increase from 1971 to 1976 in the amounts of herbicides applied to crops. Much of this increase must have resulted from the spread of conservation tillage. Conservation tillage means any of a variety of tillage practices which may differ in many details but which are distinguished from conventional tillage by three common features: (1) they rely on some instrument other than the moldboard plow to prepare the land for planting; (2) they leave enough residue from the previous crop on the soil surface to significantly reduce erosion; (3) they rely more on herbicides and less on mechanical cultivation to control weeds.

It is the third characteristic which is relevant for the present discussion. Conservation tillage relies heavily on herbicides to control weeds. In no-till systems, a polar case of conservation tillage, the

---

[22]The discussion of conservation tillage in this and following sections is based on Crosson (1981)

reliance typically is complete, that is, weed control relies exclusively
on herbicides. Other forms of conservation tillage may include some
cultivation, but all usually require more herbicides per acre for weed
control than conventional tillage. The amount more, however, is highly
variable.

Judgments about future trends in the amount of herbicides applied to
crops depend crucially upon prospects for the spread of conservation tillage relative to conventional tillage. These prospects will be shaped
in large part by the comparative economics of conservation tillage and
conventional tillage.

Economics of Conservation Tillage. Table 6-11 gives estimates of
land in conservation tillage in 1965 and subsequently, and table 6-12
shows the regional distribution in 1978 and 1979. In the period covered
by table 6-11 there were no special public programs to promote adoption
of conservation tillage. Its rapid spread since the mid-1960s, therefore, suggests that conservation tillage has important economic advantages
over conventional tillage under a wide range of soil and climate conditions.

Costs Per Acre. A survey of the literature on tillage technologies
(Crosson, 1981) indicates that total costs per acre are roughly 5 to 10
percent less for conservation tillage than for conventional tillage, the
biggest differences being for sorghum and wheat and the smallest for corn
and cotton. The difference for soybeans is intermediate. These differences reflect the net outcomes of lower costs for pre-harvest labor,
machinery, and fuel, and higher costs for pesticides, mostly herbicides.
The lower labor and fuel costs result from the elimination of some (all
in the case of no-till) cultivation to control weeds. Machinery costs
typically are lower with conservation tillage because the practice involves less disturbance of the soil, hence does not require as powerful,
and expensive, a tractor.

Conservation tillage typically requires more skilled management than
conventional tillage, but we do not expect this to seriously impede the
spread of the technology over the long term. As stated previously (Crosson, 1981, page 13), "the history of American agriculture, especially since
the end of World War II, demonstrates that American farmers can quickly

Table 6-11. Land in Conservation Tillage in the United States
(millions of acres)

Year	USDA	Percentage of harvested cropland	No-Till Farmer No-till[a]	Minimum till	Total	Percentage of harvested cropland
1965	6.6	2.3	n.a.	n.a.	n.a.	n.a.
1973	29.5	9.3	4.9	39.1	44.0	13.9
1975	35.8	10.8	6.5	49.7	56.2	17.0
1976	39.2	11.8	7.5	52.1	59.6	18.0
1977	47.5	14.1	7.3	62.7	70.0	20.7
1978	51.7	15.6	7.1	67.7	74.8	22.6
1979	55.0	16.1	6.7	78.5	85.2	24.8
1980	65.0	18.8	7.1	81.4	88.5	25.6
1981	72.2	20.5	8.7	88.1	96.8	27.4

Sources: USDA, from Gerald Darby, conservation agronomist with the Soil Conservation Service. Based on reports from SCS county field offices. No-Till Farmer (March 1981). Estimates by state agronomists of the Soil Conservation Service. Harvested cropland 1965-1980 from the U.S. Department of Agriculture, Economic Indicators of the Farm Sector: Production and Efficiency Statistics, 1980 Statistical Bulletin no. 679 (Washington, D.C., Economic Research Service, July 1982). 1981 from Thomas Frey, Economic Research Service.

[a]Defined as "where only the intermediate seed zone is prepared. Up to 25 percent of surface area could be worked. Could be no-till, till-plant, chisel plant rotary strip tillage, etc. Includes many forms of conservation tillage and mulch tillage."

n.a. = not available.

Table 6-12. Conservation Tillage by Region

Region	1980		1981	
	Millions acres	Percentage of cropland harvested	Millions acres	Percentage of cropland harvested
Northeast	2.89	21.7	3.87	28.9
Lake states	6.39	16.4	7.16	18.0
Corn Belt	27.65	31.8	28.77	33.3
Northern Plains	25.03	33.7	26.44	33.6
Appalachia	6.38	32.9	6.51	34.4
Southeast	5.54	36.0	6.00	38.7
Delta	2.67	13.6	2.85	14.3
Southern Plains	4.40	13.5	5.39	15.6
Mountain	6.63	24.7	7.28	26.2
Pacific	2.24	12.4	2.37	12.8
Total	89.82	25.96	96.64	27.4

Note: The No-Till Farmer estimates that in both 1980 and 1981 the sum of land in no-tillage, minimum tillage, and conventional tillage in the country as a whole was 300 million acres. The USDA (January 1982) shows 346 million acres of cropland harvested in 1980 and 353 million acres in 1981. If some of the "missing" cropland were in conservation tillage, the figures in the table would be too low.

Sources: Land in conservation tillage from No-Till Farmer (March 1981). Cropland harvested from U.S. Department of Agriculture, Economic Indicators of the Farm Sector: Production and Efficiency Statistics, 1980, Statistical Bulletin no. 679 (Washington, D.C., Economic Research Service, January 1982).

learn to use new, complex technologies when it is in their economic interest to do so....Indeed, the rapid spread of conservation tillage since the
mid-1960s is itself testimony to the managerial capacity of farmers and
the institutional structure which serves them....Should conservation tillage not continue to expand, it likely will be for reasons other than increased management requirements."

Yields. If there were no difference in yields between conservation
tillage and conventional tillage, the cost advantage of conservation tillage would indicate its continued rapid substitution for conventional
tillage. On erosive land conservation tillage reduces erosion 50 to 90
percent compared to conventional tillage, giving conservation tillage a
clear yield advantage for sustained cultivation of row crops on such land.
Whether the farmer would recognize the advantage and adopt conservation
tillage accordingly is much debated. For reasons given in the next chapter, we believe farmers are sensitive to the effects of erosion on yields,
even when these occur over the long term. Consequently, we believe that
on land where conservation tillage confers a long term yield advantage,
farmers will adopt it.

The question of farmers' responses to the long-term effects of erosion on yields would be moot if conservation tillage had a clear short-
term yield advantage over conventional tillage. The evidence on this is
mixed. As noted previously (Crosson, 1981, page 17), "some studies show
yields higher for conservation tillage, others show them lower and still
others show no significant differences. The specific circumstances of
soil, weather, and kind and quantity of weeds are of crucial importance in
determining relative yield response, and these conditions differ widely,
both spatially and even temporally in the same location. Other factors
less often mentioned, and apparently less important, are seed placement
and soil compaction."

The key factors affecting short-term yields appear to be soil moisture and temperature, length of growing season and weeds. There is a
strong consensus in the literature that in the plant root zone soil moisture generally is higher with conservation tillage than with conventional
tillage, and that in the spring soil temperatures are lower. Both of
these conditions have a bearing on yields of conservation tillage relative

to conventional tillage. As noted elsewhere (Crosson, 1981, page 17), "in areas where rainfall is the principal factor limiting plant growth, as in part of the Northern Plains, the moisture-conserving feature of conservation tillage is a distinct plus....The moisture-retaining characteristic of conservation tillage, however, is an advantage in any area subject to periods of drought or where soils are droughty. The characteristic is a distinct disadvantage, however, on poorly drained soils."

As the growing season shortens, the delay in spring planting imposed by the cooler soil temperatures of conservation tillage imposes an increasing yield penalty on that practice. On this score the attractiveness of conservation tillage diminishes as one moves from south to north. However, as one moves from north to south the growing season lengthens, and conservation tillage, because it saves time, opens up opportunities for double-cropping not available with conventional tillage.

Inadequate weed control is a major factor adversely affecting yields with conservation tillage. Weeds compete with crops for sunlight, moisture and nutrients, and where they cannot be controlled, yields will suffer. In some areas and for some weeds, adequate control with herbicides is difficult or impossible. Perennial weeds in particular are troublesome because available herbicides do not attack the root systems of these weeds as effectively as cultivation does.

Summary on Economics. In production of corn, sorghum, wheat, soybeans, and cotton, conservation tillage has a cost advantage of roughly 5 to 10 percent relative to conventional tillage. Crosson states (1981, p. 21), "It follows that in areas with well drained soils," long enough growing seasons, "adequate control of weeds with herbicides, and potential for double-cropping, economics would clearly favor conservation tillage." Crosson (1981), after analysis of these conditions, concludes that they are met widely enough to justify the spread of conservation tillage to 50 to 60 percent of the nation's cropland by 2010. Should there be new developments, e.g., herbicides more effective against perennial weeds, the potential spread of the technology would be increased.

Future Trends in Herbicide Use. Table 6-13 indicates how much herbicide would be used on main crops in 2010, given our projections of crop acreage, if the percentages of acres receiving herbicides and the amounts

Table 6-13.   Herbicides Applied to Crops
(million lb active ingredients)

	1976	2010	Percentage increase from 1976
Corn	207.1	269.3	30.0
Cotton	18.3	25.5	39.3
Wheat	21.9	28.0	27.9
Soybeans	81.1	179.1	121.0
Subtotal	328.4	501.9	52.8
Total of all crops	394.3		

Sources: 1976, from T.R. Eichers, P.A. Andrilenas, and T.W. Anderson, Farmers' Use of Pesticides in 1976, Agricultural Economic Report no. 418 (Washington, D.C., Economics, Statistics and Cooperatives Service, 1978). The figures for 2010 assume that the percentages of each crop receiving herbicides and the amounts applied per acre are the same as in 1976, and total acreage in each crop is as we have projected.   See text discussion for qualification of these assumptions.

applied per acre are the same as in 1976.   If conservation tillage spreads as we expect and substitutes for herbicides in conservation tillage systems are not developed, per acre applications of herbicides for corn, soybeans, and wheat will rise from 1976 levels.   The continued spread of conservation tillage based on currently available herbicides therefore implies larger applications of herbicides in 2010 than shown in table 6-13.   We have no basis for estimating the additional amount, but it could be substantial, particularly if the percentage of wheat land receiving herbicides is increased.   (Most corn, soybean, and cotton land already is treated with herbicides--see table 6-6.)

References

Aldrich, S. 1980. Nitrogen in Relation to Food, Environment and Energy (Urbana-Champaign, Agricultural Experiment Station, University of Illinois).

Andrilenas, P. 1974. Farmers' Use of Pesticides in 1971 - Quantities, Agricultural Economic Report no. 252 (Washington, D.C., Economic Research Service).

Crosson, Pierre. 1971. Conservation Tillage and Conventional Tillage: A Comparative Assessment (Ankeny, Iowa, The Soil Conservation Society of America).

Douglas, John. 1978. Remarks given at the Third World Fertilizer Conference, San Francisco, Calif.

Eichers, T. R., and P. A. Andrilenas, and T. W. Anderson. 1978. Farmers' Use of Pesticides in 1976, Agricultural Economic Report no. 418 (Washington, D.C., Economics, Statistics Cooperative Service).

Fertilizer Institute. 1976. The Fertilizer Handbook (Washington, D.C.).

Heady, Earl. 1982. "The Adequacy of Agricultural Land: A Demand-Supply Perspective," in Pierre Crosson (ed.), The Cropland Crisis: Myth or Reality (Baltimore, Md., Johns Hopkins University Press for Resources for the Future).

Headley, J.C. 1981. Pest Control as a Production Constraint for Grain Crops and Soybeans in the United States to 1990, Research Bulletin 1038 (Columbia, Missouri Agricultural Experiment Station).

Hemphill, Gregory. 1980. "Fertilizer Use: Alternative Management Practices and Nutrient Materials" (Washington, D.C., Resources for the Future).

Lockeretz, W., G. Shearer, and D.H. Kohl. 1981. "Organic Farming in the Corn Belt," Science vol. 211 (Feb.) pp. 540-547.

Miranowski, J. 1979. "Integrated Pest Management in Corn Rootworm Control: A Preliminary Technology Assessment," paper given at the annual meeting of the American Agricultural Economics Association, Pullman, Wash., July 29-August 1, 1979.

National Academy of Sciences. 1975. Pest Control: An Assessment of Present and Alternative Technologies vol. III, Cotton Pest Control (Washington, D.C., NAS).

No-Till Farmer. 1981. (March).

Office of Technology Assessment (OTA). 1980. Pest Management Strategies, vol. II Working Papers (Washington, D.C., GPO).

Pimentel, D., C. A. Shoemaker, E. L. LaDue, R. B. Rovinsky, and N. P. Russell. 1977. Alternatives for Reducing Insecticides on Cotton and Corn: Economic and Environmental Impact (Washington, D.C., EPA).

Stanford, G. 1978. "Prospects for Using Nitrogen More Effectively in Crop Production," paper given at the 1978 meeting of the Northeastern Branch, American Society of Agronomy, University of Connecticut, Storrs.

USDA. 1971. Cropping Practices: Corn, Cotton, Soybeans, Wheat, 1964-70, SRS-17 (Washington, D.C., Statistical Reporting Service).

_____. 1977. 1977 Fertilizer Situation, FS-5 (Washington, D.C., Economic Research Service, January).

_____. 1979. Agricultural Statistics (Washington, D.C., GPO).

_____. 1980. 1981 Fertilizer Situation, FS-10 (Washington, D.C., Economics, Statistics and Cooperatives Service, December).

_____. 1980. Crop Production 1979 Annual Summary, CrPr 2-1 (Washington, D.C., Crop Reporting Board, January 15).

_____. 1980. Report and Recommendations on Organic Farming, prepared by a USDA study team on organic farming (Washington, D.C., USDA, July).

_____. 1982. Economic Indicators of the Farm Sector: Production and Efficiency Statistics, 1980, Statistical Bulletin no. 679 (Washington, D.C., Economic Research Service, January).

University of Illinois. 1979. Tillage Systems for Illinois, College of Agriculture Cooperative Extension Service Circular 1172 (Urbana-Champaign, University of Illinois).

von Rumker, R., E. W. Lawless, and A. F. Meines; with K. A. Lawrence, G. L. Kelso, and F. Horay. 1975. Production, Distribution, Use and Environmental Impact Potential of Selected Pesticides (Washington, D.C., EPA Office of Pesticide Programs and the Council on Environmental Quality).

Chapter 7

ENVIRONMENTAL IMPACTS OF PROJECTED PRODUCTION
AND RESOURCE USE

Introduction

In the scenario projected to 2010 the increases in crop acreage and
in amounts of fertilizers and pesticides used imply, in our judgment, not
only rising economic costs of production. They also suggest increasing
environmental costs. The environmental costs considered are those resulting from increased use of fertilizers and pesticides, from the expansion
of irrigation, and from erosion.[1] Comprehensive and generally acceptable
quantitative estimates of these costs are not available, and perhaps not
possible, even theoretically. One reason is lack of basic knowledge,
e.g., the contribution of phosphorus fertilizer to eutrophication of water
bodies compared to that of municipal wastes, or the fate of pesticides
after they are applied to farmers' fields. A second major reason is the
conceptual difficulty of valuing some of these impacts, e.g., human illnesses or deaths from pesticides.

For these reasons our assessment of environmental costs makes no
attempt at quantification in the sense of attempting to estimate the
dollar value of the costs. Instead we seek to make judgments on two
questions: (1) are the environmental impacts of crop production presently
of such a nature as to require new policies or modification of existing
policies to deal adequately with them; and (2) given our projected in-

---

[1]Animal wastes are an important source of pollution of both ground
and surface water. Much of the waste, however, is from large feed lots,
now subject to control by the EPA as so-called point sources of pollution. Moreover, the major expansion in agriculture, hence the major
source of increased environmental damage, will be from crop production,
not animal production.

creases in resource use, are the future impacts likely to be more or less severe than at present? These judgments set the stage for the discussion in the following chapter of policy issues and options.

Fertilizer

Nature of the Environmental Impacts

The principal concerns about fertilizers in the environment have to do with effects of nitrogen in water on human and animal health and of nitrogen and phosphorus in accelerating eutrophication of water bodies by stimulating growth of aquatic plants.[2]

Nitrates in water may constitute a hazard to both humans and animals that drink the water, not because the nitrites themselves are particularly toxic but because of their potential for conversion to nitrates, which can be when present in sufficient concentration. Babies are more likely to suffer from nitrite poisoning than adults. Deaths and illnesses among farm animals also have been attributed to nitrite intoxication (Luhrs, 1973). Nitrogen gas dissolved in water or nitrogen in ammonia may kill or injure fish (Patrick, 1973).

Eutrophication is a process involving the nutrient enrichment of lakes and reservoirs, the resultant growth of plant life, and the subsequent decline in the water's dissolved oxygen supply because of the oxygen demand of decaying plants. If severe, eutrophication makes the water incapable of supporting fish and spoils its value for some kinds of recreation. Eutrophication is a natural process, but it can be accelerated

---

[2]Denitrification of nitrate-N releases nitrogen oxides, one of which, nitrous oxide (N_2O), may attack the earth's ozone shield. The resulting increase in solar radiation reaching the earth's surface would increase the risk of skin cancer. This possibility received considerable popular attention in the mid-1970s, particularly after press reports of work done by Michael McElroy at Harvard. According to Aldrich (1980, pp. 222-223), a subsequent study by the National Academy of Sciences indicated a lag of about 100 years between the application of nitrogen fertilizer and effect upon the ozone shield. Moreover, even a very large increase worldwide in nitrogen fertilizer use would reduce the ozone shield only 1.5 to 3.5 percent by 2100. Perhaps for these reasons the issue has faded from public discussion. We do not consider it further in this report.

by the addition of nutrients supplied by man, including nitrogen and phosphorus carried by run-off water and soil eroded from farmland. The amounts of these nutrients may be very small yet have significant effects in stimulating growth of aquatic plants. According to Holt, Johnson, and McDowell (1973), as little as 10 parts per billion (ppb) of phosphorous and 100 ppb of nitrogen are enough to support growth of undesirable amounts of aquatic plants.

Present Severity of the Problems

Some years ago one of the authors assessed the severity of the nitrate problem in the United States and agriculture's contribution to it in the following words (Crosson and Frederick, 1977, pp. 192-195):

> The evidence about the present seriousness of the environmental hazards of fertilizers is not easy to interpret. Numerous studies show the presence of nitrates in groundwater, for example, in parts of Connecticut, Long Island, Illinois, the High Plains, Minnesota, and California (Martin, Fenster and Hanson, 1970; Miller, Deluca, and Tessier, 1974). In some of these places the nitrate-N level has exceeded ten parts per million, the standard for drinking water set by the Public Health Service. It is not clear, however, that fertilizer used in agriculture is the principal source of the nitrate in these waters. Martin and coauthors cite a study by Smith of 6,000 rural water supplies in Missouri which concluded that animal wastes and septic tank drainage were the principal sources of nitrates. Studies in Minnesota show that rural wells were contaminated by nitrate long before fertilizer use become important, and that between 1899 and 1963 change in the quality of water in major aquifers was minor. A study of an area near San Luis Obispo, California, showed that nitrate in groundwater "...was substantial and mostly associated with the native soil organic matter complex supplemented with sewage waste from area homes and to a lesser extent from lawn and farm fertilization" (Miller, DeLuca and Tessier, 1974, p. 309). In their study of groundwater pollution in the northeastern states Miller and coauthors assigned a "moderate" priority to fertilizer for additional research and control measures while giving high priority to domestic, municipal, and most other sources of wastes.

> The extent of fertilizer pollution of surface waters in the United States also is unclear, although the problem seems to have been less investigated. In a study of the southern High Plains Goldberg concluded that "nitrogen fertilizer applied to the farmland adds little nitrate to the surface water" (Goldberg, 1970). Aldrich says that the nitrate content of rivers in the central Corn Belt for which long-term records are available shows an upward

trend. The nitrate content of the Mississippi River at the point where it leaves the Corn Belt about doubled from the mid-1950s to the mid-1970s but was still less than the maximum acceptable amount set by the U.S. Public Health Service for drinking water. The nitrate content of the Illinois River increased about one-quarter in this period but still was only about one-half of the Public Health Service's standard, while the nitrate in the Missouri declined about 50 percent between 1950 and 1970. In some small creeks and rivers in Illinois, however, the nitrate content sometimes exceeds the standard (Aldrich, 1976).

The National Academy of Sciences cites a report on fertilizer pollution in Illinois by the Illinois Pollution Control Board (IPCB) which concluded that "there is no factual basis for imposing restrictions on the use of fertilizer at this time." The conclusion applied both to nitrogen and phosphate fertilizers....The EPA subsequently issued a policy statement in general agreement with the decision of the IPCB regarding fertilizer, but it also stressed the need for additional information (National Academy of Sciences, 1975b, pp. 92-93).

In our reading of the evidence on the environmental threats of fertilizer, two important aspects stand out. One is that human deaths or illnesses attributable to nitrite poisoning are extremely rare. Animal poisonings are more common, but where caused by drinking water the nitrate content of the water was seven to fifteen times the standard set by the U.S. Public Health Service (Aldrich, 1976). Enforcement of the standard should eliminate such instances. Fish kills or injuries attributable to forms of nitrogen in surface water also are of small overall importance.

The second aspect is that much of the nitrogen and phosphorus found in water bodies comes from municipal and industrial discharges and from the natural leaching of these nutrients already in the soil.

Consideration of these two aspects of the evidence leads us to conclude that with present levels of use fertilizers do not pose threats to the environment so severe that changes in present policies are required to deal with them.

We believe this assessment of the nitrate problem is still valid. After a thorough review of the data on nitrates in surface waters, Aldrich (1980, p. 141) concludes: "The evidence is convincing that, based upon available data, the nitrate issue in surface waters in the United States in relation to health is limited mainly to the Midwest and to certain rivers in intensively farmed regions of California."

Aldrich's review shows that even in the Midwest nitrate concentrations in rivers only occasionally exceed the 10 ppm nitrate-N concentration standard set by the U.S. Public Health Service.

Nor have we found anything in the more recent literature to alter our judgment about the contribution of nitrogen fertilizers to contamination of groundwater. The Council on Environmental Quality (CEQ, 1979) asserted that surface and subsurface disposal of wastes are among the major sources of groundwater contamination in the country. Disposal of industrial wastes in this manner is mentioned in particular. In this context, the CEQ (1979, pp. 110-111) notes that infiltration of nitrates from nitrogen fertilizers is another important source of groundwater contamination. However, earlier in the same report (CEQ, 1979, pp. 107-109), a table summarizing groundwater quality conditions in 18 river basins making up the 48 contiguous states is presented which mentions fertilizer contamination in only two regions, the Souris-Red-Rainey (northern Minnesota and North Dakota) and the Pacific Northwest. Irrigation return flow is mentioned as a contaminant of groundwater in the Great Basin, and this presumably includes nitrates from fertilizer, although the CEQ report does not say so explicitly. The table conveys the strong impression, however, that with the exception of local "hotspots" in the regions mentioned, contamination of groundwater by nitrogen fertilizer is not presently a major problem.

This interpretation is supported by a more recent publication by the CEQ (1981) dealing with contamination of groundwater. The CEQ cites an EPA report (1977b) as identifying the disposal of industrial wastes at industrial impoundments and solid waste disposal sites as the most important source of groundwater contamination. Septic tanks, municipal waste water, mining and petroleum exploration and production are listed as of secondary importance. Agriculture is not mentioned as of even secondary importance as a source of groundwater pollution.

A study of the Santa Maria valley in California in 1976 showed that 39 percent of the nitrogen fertilizer applied to 57,000 acres of vegetable, field and fruit crops was leached below the root zone, but there is no indication that nitrate concentration in groundwater in the area exceeded 10 ppm as a consequence (Lund and coauthors, 1978). A study of the Upper

Santa Ana River basin of California in the early 1970s revealed several
"hotspots" where nitrate concentrations were high (Ayers, 1978). In the
dairy farm area of the Chino Basin, a small part of the whole basin, nitrate-N concentrations at the top of the water table underlying cropland
sites averaged 45 ppm. Deep well water in the same vicinity averaged 6
ppm. Ayers infers from these numbers that the dairy wastes had not yet
had full impact on nitrate-N concentrations in the groundwater. He evidently believes that animals rather than fertilizers are the main contri-
butors to the build-up of nitrate-N in the groundwater.

A problem area that has emerged since our previous review of the
literature on fertilizer pollution (or which our review missed) is the
Sandhills region of Nebraska. In irrigated parts of that region nitrate-
N concentrations in groundwater have risen to 20 ppm. With good irrigation management concentrations should not rise above 20 to 25 ppm, but
with continued expansion of irrigation in the area they are not likely to
fall below those levels.[3]

Eutrophication. The Environmental Protection Agency conducted a
nationwide eutrophication survey between 1972 and 1977 which found that
two-thirds of the 800 lakes studied were eutrophic and another 4 percent
were hypereutrophic.[4] On its face this would seem to suggest that eutrophication is a major national problem. It is not so treated in the cited
CEQ report, however, nor in recent annual reports to Congress by the EPA
on the condition of national water quality. This is not to say that the
CEQ and the EPA reports dismiss eutrophication as a national problem.
They do not. But the reports convey no sense that the problem is of the
magnitude suggested by the finding of the EPA's eutrophication study that
over 70 percent of the lakes surveyed were eutrophic or hypereutrophic.
The reason, perhaps is that eutrophication is a matter of degree, and a
lake classed as eutrophic under the EPA definition still can be fishable
and swimmable and provide other recreational values.

---

[3]Darryl Watts, University of Nebraska, cited in Frederick (1982).

[4]Council on Environmental Quality (1979, p. 92). According to the
CEQ, a eutrophic lake usually has murky, greenish water and measurable
amounts of plant productivity, and a hypereutrophic lake has very murky
water and extremely high levels of plant biomass.

Whatever the seriousness of the eutrophication problem, fertilizers apparently contribute less to it in most instances than other sources of nutrients. Phosphorus is usually the nutrient limiting plant growth in lakes and reservoirs, and municipal, industrial, and other non-agricultural wastes typically contribute more phosphorus (and other nutrients) to these water bodies than fertilizers applied in agriculture (CEQ, 1979, p. 95). In discussing eutrophication in Lakes Erie and Ontario, the most seriously affected of the five Great Lakes, the CEQ report puts high loads of nutrients from municipal sewage ahead of those from agricultural and urban runoff as the principal cause. According to the EPA (1977a, p. 9), control of phosphorous discharges from municipal waste treatment plants is expected to have a major effect in reducing eutrophication in heavily populated regions.

Future Severity of the Problems

We have projected an increase of 87 percent from 1977/79 to 2010 in nitrogen applied for all purposes. For phosphorus the increase is 68 percent (see table 6-3).[5]

The projected increases of fertilizers applied to the four main crops (corn, wheat, soybeans, and cotton) from 1977/79 to 2010 are 61 percent for nitrogen and 71 percent for phosphorus. Seventy to 75 percent of the increase in nitrogen applied to these crops and 80 percent of the increase in phosphorus is attributable to the increased acreage in corn, wheat, and soybeans.[6] The percentage increases for nitrogen are less for the four main crops than for total applications because "all other" uses of nitrogen are projected to increase faster than uses on the four main crops. "All other" uses of fertilizer includes that put on crops other than the main four, as well as that used on gold courses, parks, home gardens, and lawns, and so on.

---

[5]There is little concern about the environmental impacts of potassium fertilizers, and we do not consider them here.

[6]We project a small decline in fertilizer applied to cotton by 2010 so we do not consider fertilizer use on cotton in this discussion (see table 6-3).

Our projections in chapter 6 of land in crops assume that in 2010 lands in crops other than main crops will be 16 million acres less than in 1977, reflecting a 10 million acre decline in land in hay and a 6 million acre decline in land in sorghum, oats and barley. We have not projected fertilizer applications per acre of land in "other crops", but we do not expect much increase, for the same reasons that little increase is expected in per acre applications to main crops. It follows that most of the projected increase in "all other" uses of fertilizer is for non-crop purposes. Since we are interested in this study in the environmental impacts of crop production, we give no further attention to "all other" uses of fertilizer. We note, however, that should concern about environmental impacts of fertilizers mount in the future the "all other" uses likely will require special attention.

The environmental impacts of the projected increases in fertilizers used on the four main crops depend on how much of the additional nutrients move to ground and surface water and on existing nutrient concentrations in the water. For the moment we focus on movement of the additional nutrients.

Movement of Nutrients. For several reasons we expect the increase in movement of nutrients to water bodies to be proportionally less than the increase in amounts of fertilizers applied. As noted above, 70 to 80 percent of the increase in amounts applied of both nitrogen and phosphorus from 1977/79 to 2010 is because of the expansion of land in corn, wheat, and soybeans. About 16 million acres of the additional land is now in hay (10 million acres), sorghum, oats, and barley (6 million acres). We have no completely reliable information on how much fertilizer currently is applied to these 16 million acres, but several sources suggest a conservative range of 50 to 75 pounds per acre of both nitrogen and phosphorus (International Minerals and Chemical Corporation, n.d.). If we assume 60 pounds, then the 16 million acres currently receive 960 million pounds of nitrogen and phosphorus per year. Consequently, the net increases by 2010 in the amounts of these materials applied to land in corn, wheat, and soybeans would be 960 million pounds less than the gross amounts we have projected. In percentage terms, the net projected increase for nitrogen would be 53 percent compared with a gross increase of

66 percent. For phosphorus the net increase would be 51 percent compared with a gross increase of 75 percent.

The increase in movement of nutrients may not be proportional to the net increases in amounts applied also because fertilizer technologies are likely to become more efficient in response to higher prices, as discussed in the previous chapter, and because of the expected increase in conservation tillage relative to conventional tillage.

Effects of Increased Efficiency. The anticipated adoption of fertilizer materials and practices which reduce nutrient losses would tend to hold the proportionate increase in movement of nutrients below the increases in total amounts applied. We have no solid basis, however, for estimating how much losses might be reduced. Stanford (see reference to chapter 6) cites a study in Nebraska indicating that losses of sidedressed nitrogen were 30 percent less than losses of N applied in the fall or spring. Stanford goes on to assert that an "appreciable" increase in nitrogen efficiency can be achieved by more realistic recommendations of application rates and proper timing of application, but he gives no numbers.

Nitrogen fertilizer losses vary widely from place to place, depending upon soil moisture and temperature and other factors. According to Hinish (1980), losses can range from 20 to 60 percent. For illustrative purposes, we assume average nitrogen losses of 40 percent and that improvements in practices and materials reduce losses by 20 percent between 1977/79 and 2010. With these assumptions, the projected 53 percent net increase in nitrogen applied to corn, wheat, and soybeans would result in a little more than 20 percent increase in nitrogen losses.[7]

Aldrich (1980, p. 119) expects use of nitrogen fertilizer in the Corn Belt to increase. He argues, however, that higher prices, among other factors, will induce farmers to use fertilizer more efficiently.

---

[7]For example:

	1977/79	2010
Amount of nitrogen applied	100	153
Percentage lost	40	32
Amount lost	40	49

The amount lost would increase a little more than 20 percent.

Indeed, he asserts that the increase in amount of nitrogen applied "...will closely follow the rise in efficient utilization by crops."

Losses of phosphorous fertilizers range from 60 to 80 percent (Hinish, 1980). Assuming an average of 70 percent and that losses fall by 20 percent by 2010, the projected 51 percent net increase in phosphorous applied would result in about a 20 percent increase in losses.

Effects of Tillage Technologies. Losses of fertilizers applied to land in the three main crops in 2010 also will be influenced by the kind of tillage technologies used on this land. We concluded in the previous chapter that the economics of conservation tillage would favor its spread to 50 to 60 percent of the nation's cropland by 2010 (increasing its share two to three times). Most conservation tillage is on corn, soybean, and wheat land, so we anticipate conservation tillage of land in these crops to increase significantly by 2010.

The effects of the spread of conservation tillage on movement of nutrients to water bodies is uncertain. It will be shown below that conservation tillage greatly reduces erosion compared to conventional tillage, and a number of studies have demonstrated that most of the nitrogen and phosphorous moved from farmers' fields to surface water bodies is carried by eroded soil.[8] The nitrogen so moved is in the organic form and when deposited with sediment in rivers, lakes, and reservoirs, is mineralized to the nitrate form only very slowly. Since it is nitrate-N which may adversely affect water quality, the advantage of conservation tillage in reducing erosion, and hence losses of organic nitrogen, does not significantly reduce the impact of nitrogen on water quality.

Wauchope, McDowell, and Hagen (n.d.) cite a number of studies indicating that, unlike the nitrogen carried by eroded soil, 5 to 40 percent of the phosphorous so carried is available to support aquatic plant growth.

Nutrients are carried to water bodies by runoff as well as by eroded soil, and there is much evidence that nutrient concentrations in run-off from conservation tilled fields are higher than in runoff from conventionally tilled fields (Wauchope, McDowell and Hagen, n.d.; Barisas and coauthors, 1978). However, the amount of runoff typically is less from conser-

---

[8]Barisas and coauthors (1978) and studies cited therein.

vation tilled fields because the crop residues and rougher soil surface found with conservation tillage reduce the velocity of water flow and increase infiltration. The net outcome of these opposing effects in delivering nutrients to receiving waters is uncertain. It depends upon the specific differences between conservation tillage and conventional tillage with respect to nutrient concentrations in runoff and amount of runoff, and these differences follow no regular pattern. The reduction in runoff water with conservation tillage may result in more infiltration of nitrate-N to groundwater than with conventional tillage. Unlike nitrate-N, phosphorus is strongly adsorbed by soil and so does not leach to groundwater in significant amounts.

The discussion suggests that conservation tillage may pose a greater threat to groundwater quality than conventional tillage because of increased leaching of nitrate-N, but that the comparative effects on delivery of nitrate-N to surface waters is too dependent on specific local conditions to warrant a general conclusion.

The effect of the spread of conservation tillage on the threat of phosphorus to water quality also is not clear. The outcome depends upon the reduction in available phosphorus carried by eroded soil relative to the increase carried by runoff water. In general, the advantage of conservation tillage with respect to the total amount of available phosphorus delivered to water bodies will be greater (1) the greater the reduction in erosion; (2) the smaller the difference between conservation and conventional tillage in concentration of total P in sediment; (3) the greater the ratio of available P to total P in sediment; (4) the higher the sediment delivery ratio; (5) the smaller the difference between the two tillage systems in concentration of phosphorus in runoff water. While erosion typically will be substantially less with conservation tillage and the concentration of available P in both sediment and runoff water higher, the range of difference in these and the other variables is large. Consequently, no general conclusion about the effect of the spread of conservation tillage on the delivery of phosphorus to water bodies seems warranted.

While we cannot say how the shift to conservation tillage would affect delivery of phosphorus and nitrate-N to surface waters, it seems reasonably well established that the shift would increase movement of nitrates to groundwater. Thus the shift probably would offset to some extent the effect of more efficient materials and practices in reducing nitrate losses. We lack data to precisely estimate the net outcome, but some plausible numbers suggest that the increase in nitrate losses could be about 30 percent. This would be the case if conservation tillage of land in corn, wheat and soybeans increased from 30 percent in 1977/79 to 60 percent in 2010, if nitrate losses were 50 percent from conservation tilled land and 40 percent from conventionally tilled land, and if losses with both kinds of tillage were 20 percent less in 2010 than in 1977/79.[9]

We conclude that from 1977/79 to 2010 losses of nitrogen fertilizer from land in corn, wheat, and soybeans will increase significantly less than the projected 66 percent gross increase in amounts applied. The increase in phosphorus losses also is likely to be significantly less than the gross projected increase of 75 percent, although this outcome is clouded by uncertainty about the effect of the shift to conservation tillage on phosphorus losses.

Whether these increases in nutrient losses would cause water quality problems depends upon nutrient concentrations in the receiving waters. Since existing concentrations of nitrate-nitrogen do not appear to pose serious problems except, perhaps, in a few local "hot spots" around the country, some increase in concentrations evidently could be accommodated without serious consequence. Whether an increase on the order of 30 percent would be cause for concern is uncertain, but we consider it unlikely.

Our earlier discussion indicated that municipal wastes are a major source of phosphorus in water and that significant reductions in this source are expected as a result of increased investment in municipal

---

[9]For example:

	1977/79			2010		
	Conservation tillage	Conventional tillage	Combined amount	Conservation tillage	Conventional tillage	Combined amount
Amount of nitrogen applied	30	70	100	91.8	61.8	153
Loss percentage	50	40	43	40	32	38.4
Amount lost	15	28	43	36.7	19.6	56.3

Total losses would increase about 30 percent.

waste treatment around the country. This suggests that the increased losses of phosphorus from land in corn, wheat, and soybeans would not cause major increased damage to water quality. Accordingly, we do not further consider phosphorus losses.

The estimates so far presented of fertilizer applications and losses are for the nation as a whole. They mask widely varying regional increases in application and losses reflecting primarily regional differences in projections of land in the three main crops. The smallest relative increases in nitrogen applied--all about 40 percent--are in the Lake States, Corn Belt, and Northern Plains. The largest percentage increases in nitrogen applied are in the Mountain States (280 percent), Southeast (195 percent), Southern Plains (180 percent), Pacific (150 percent), and Delta (100 percent).

We expect that in both the Corn Belt and Northern Plains about 2 million acres of land now in sorghum, oats, and barley will be converted to corn, wheat, and soybeans by 2010. Consequently, the net increases in nitrogen applied in these regions would be somewhat less than the projected 40 percent gross increase. Allowing for more efficient nitrogen materials and practices and for the spread of conservation tillage, nitrate losses in these regions, and in the Lake States, might increase 20 to 25 percent by 2010. Increases of this magnitude perhaps would present few problems in the Lake States and Corn Belt, where nitrate concentrations in water evidently are not now serious. The situation might be different, however, in the Northern Plains. As noted earlier, nitrate concentrations in groundwater in some parts of Nebraska now are about 20 ppm, twice the Public Health Service's standard for safe water. In such areas, the prospect of increased nitrate losses of 20 to 25 percent indicate cause for concern.

Even allowing for the spread of more efficient nitrogen materials and practices, the projections of nitrogen applied in the South, Southwest, Mountain Region and Pacific Coast imply substantial percentage increases in nitrate losses in these regions. Without information about present nitrate concentrations in regional water bodies, there is no basis for judging whether the increased losses would pose water quality problems. In our judgment, however, the increases are sufficiently large to

indicate that such problems may emerge by 2010. The Southeast in particular would bear watching. Much of the expansion in crop production and
fertilizer use in that region will be with irrigation on the sandy soils
of Georgia and northwest Florida. The potential for increased leaching
of nitrate-N to the groundwaters of the region looks large.

Pesticides

Nature of the Environmental Impacts

The presence of pesticides in the environment arouses concern because
often these materials act not only against the target pests but also
against other organisms. They may pose threats to human health and reproductive capacity; do damage to non-target species of plants, insects, soil
and water microorganisms, and wildlife; and cause the build-up in pests,
especially insects, of genetic resistance to pesticides.

Present Severity of the Problems

Pimental and coauthors (1980) have estimated the total environmental
costs of pesticide use in the United States at $839 million annually. This
includes 7 categories of environmental cost ranging from reduced natural
enemies and increased genetic resistance to pesticides ($287 million) to
fish and wildlife losses ($11 million). Human pesticide poisonings, in-
cluding those that result in fatalities, are included ($184 million).

Pimental and coauthors also state that they were unable to estimate
some important classes of environmental costs of pesticides. Had these
been included, total costs would have been "several times" higher than
those reported.

Total direct and environmental costs of pesticides are estimated by
Pimental and coauthors (1980) to be about $3.35 billion and total benefits
to be $10.9 billion, benefits thus exceeding total costs by a factor of 3.
If environmental costs are in fact 3 times as high as those estimated by
Pimental, benefits exceed total costs by a factor of 2.

Pimental and his coauthors are aware of the tenuous underpinnings of
these estimates--indeed they emphasize them. We present the estimates here
because to our knowledge they are the only ones available, and they are

made by reputable scientists. Our presentation of the estimates does not
imply endorsement of them. In fact, we are skeptical that reliable estimates of environmental costs of pesticides are possible at present.

There are two reasons for our skepticism. One is the conceptual difficulty of valuing some of these damages, e.g., human deaths or genetic
abnormalities, even in cases where the damages clearly are attributable to
pesticides. Pimental and coauthors (1980) assign a value of $1 million
per life to the fatalities from pesticide poisoning, but they acknowledge
that there is no really satisfactory way of valuing these losses.

The second reason for our skepticism is that little is known about
the paths by which pesticides move through the environment and their ultimate fate, making it extraordinarily difficult to identify, let alone assign values to the damages they do. In the literature on environmental
impacts of pesticides, this second difficulty receives far more attention
than the conceptual problem. For example, von Rumker and coauthors (1975,
p. 96) cite two comprehensive reports on environmental impacts of pesticides (Mrak, 1969; and Pimentel, 1971) as indicating

> ...that relatively little information is available on the toxicity
> and hazards of pesticides and their residues to nontarget organisms
> under field conditions. Information is especially sparse concern-
> ing the effect, if any, of pesticide residues on lower aquatic and
> terrestrial organisms. There is a copious literature on the effects
> of individual pesticides on isolated organisms or systems in the
> laboratory or greenhouse over short periods of time. However,
> such studies are usually far removed from field conditions, and
> their results do not answer the question of their significance
> in regard to field conditions.
>
> Pesticide monitoring studies have centered on chlorinated
> hydrocarbon insecticides. The possibility that pesticides may
> affect terrestrial or aquatic ecosystems in other ways [than through
> bioaccumulation and biomagnification] has received less attention.
> By virtue of their physical, chemical, and biological properties,
> fungicides and herbicides are more likely to affect the lower tro-
> phic levels of food chains. It is not known whether or not cur-
> rently practiced monitoring and observation methods would detect
> such effects prior to the occurrence of massive ecological damage.

In a more recent review of the literature dealing with the effects
of pesticides on water quality, Wauchope (1978) concludes that, at least
in principle, reasonable estimates can be made of edge-of-field losses of

pesticides to rivers and lakes; however, information is most needed about the fate of pesticides after they leave the field. Wauchope argues (1978, p. 471) that

> ...Overall assessments of <u>runoff</u> impact must include judgments on such factors as the time and distance of impact of a given field runoff event and the ability of local ecosystems to recover from temporary high concentrations of a pesticide. The dynamics of dilution and sediment exchange, and uptake, transfer, and metabolism by aquatic life of most of the pesticides presently in use are not known. Without this knowledge, the impact of a given pesticide input or the quality of water in a river or lake cannot be predicted.

A convincing assessment of the current severity of the environmental impacts of pesticides thus is not possible. Consideration of some of the skimpy evidence about these impacts nonetheless is appropriate. We deal with insectides and herbicides. These accounted for 86 percent of the pesticides applied to crops in 1976 (Eichers, Andrilenas, and Anderson, 1978). Fungicides, the principal other class of pesticides, are nontoxic or only slightly toxic to mammals. Consequently, in concentrating on insecticides and herbicides we miss little of significance concerning environmental impacts.

<u>Insecticides</u>. The initial concern with insecticides was primarily with the effects of the organochlorine compounds, principally DDT and similar persistent materials, on wildlife and humans. While generally not highly toxic to vertebrates, the tendency of these materials to persist in the environment and to increasingly concentrate in body tissue at higher levels in the food chain made them suspect. When tests on laboratory animals indicated that, in high enough dosages, some of the organochlorine compounds were carcinogenic and/or teratogenic,[10] concern mounted still higher. Eventually, beginning with DDT in 1972, the EPA banned or tightly restricted the use of the organochlorine compounds, and by 1976 the only one still in general use was toxaphene. However, because of their persistence, traces of these materials still are lurking about in the environment. To the extent that they pose a threat, the threat remains, although diminishing with time. If the threat includes cancer in

---

[10]Teratogenic materials cause fetal malformations.

humans--this is suspected but not conclusively demonstrated--the ultimate
signs of it may not disappear for decades because of the long latency
period of the disease.

We noted in the previous chapter that while the use of organochlorine
insecticides has declined sharply since the early 1970s, use of organophosphorus and carbamate compounds increased enough to more than offset
the decline. By comparison with the organochlorines, these compounds are
not persistent nor do they bioaccumulate. Unlike the organochlorines,
however, many of the organophosphorus and carbamate compounds are highly
toxic to humans and other nontarget organisms. Consequently, the nature
of the threat of these compounds is quite different from that of the
organochlorines. Damages inflicted by the latter typically are subtle,
diffused widely, both geographically and among affected individuals, and
long-term. Damages of the organophosphorus and carbamate compounds typically are sharp, localized, and short-term. They in fact have many of the
characteristics of industrial accidents rather than characteristics we
typically associate with environmental impacts.

According to Pimentel and coauthors (1980), there were fifty-two accidental deaths from pesticide poisonings in 1974, a significant decline
over the preceding two decades. The number of intentional deaths (sui-
cides and homicides) was about three times the number of accidental deaths.
The total number of human poisonings from pesticides was estimated by Pimentel and coauthors to be about 45,000 per year. These were attributable
to all pesticides, not just organophosphorus and carbamate insecticides.

One of the principal concerns with insecticides is the tendency for
insects to build genetic resistance to these materials. When this happens
farmers find themselves increasingly unable to control the resistent
insect and crop losses mount. Increasing amounts of insecticide are applied, costs rise, damages to beneficial insects and other nontarget
organisms increase, and resistance in the target insect is strengthened
even more. Ultimately, the insecticide may become completely useless,
and the farmer must fall back on a substitute, if one is available. Perhaps the outstanding example of the buildup of genetic resistance was the
boll weevil's response to DDT. By the 1960s the insect had become so
resistant that the use of DDT had begun to decline sharply some years

before the EPA's action to ban it. The tobacco budworm and cotton boll worm also became resistant to DDT and subsequently to some of the more important organophosphorus compounds. Indeed, the increasing resistance of these insects gave major stimulus to development of nonchemical means of control of cotton insect pests in Texas, described in the previous chapter.

Since the mid-1970s, use of synthetic pyrethroids to control cotton insect pests, particularly the tobacco budworm, has spread widely. The pyrethroids have low persistence--indeed their rapid rate of degradation may weaken their effectiveness--and low toxicity to mammals. They may be highly toxic to fish, but they are used in such small amounts and degrade so rapidly that the probability of a significant threat to aquatic life is small. The increasing adoption of synthetic pyrethroids by cotton farmers, therefore, has tended to ease the total impact of insecticides on the environment, especially to the extent that the pyrethroids have substituted for the highly toxic organophosphorus compounds.

Herbicides.[11] Most herbicides have low toxicity to people. Paraquat, widely used with conservation tillage, is an exception. There is some evidence that a number of herbicides, including paraquat, may be carcinogenic or mutagenic. Giere, Johnson, and Perkins (1980) cite two studies, one of which raises a question about paraquat as a mutagen, and another which failed to find mutagenic activity. Chemical and Engineering News (1980, p. 4) refers to studies in Europe indicating that 2,4,5-T and 2,4-D increase the risk of of developing certain kinds of cancer. Compound 2,4, 5-T, the "agent orange" herbicide used in Vietnam for defoliation, is thought by some to also produce birth defects and other reproductive abnormalities. However, the aforementioned article also reports a study, by the U.S. Government's National Toxicology Program, of the effects of 2,4,5-T and 2,4-D on male mice which showed no significant effect on fertility, reproduction, germ cell toxicology, or survival and development of the offspring of the exposed animals.

Atrazine, which accounts for almost 25 percent of all herbicides applied to crops in the United States, most of it on corn, has low toxi-

---

[11]This section is taken from Crosson (1981).

city to humans. There is evidence, however, that atrazine may be transformed metabolically by plants to form a substance which is mutagenic
(Plewa and Gentile, 1976). Giere, Johnson, and Perkins assert further that
atrazine can be transformed in the human stomach to a nitrogen derivative
under strong suspicion as a carcinogen.

A study by M.J.S. Hsia of propanil, another low toxicity herbicide
used on rice, showed that in the soil propanil is metabolized by first
fungi and then microorganisms to a compound very similar to dioxin, the
teratogen found in Agent Orange (Science News, 1973).

Concern also has been expressed that extended application of herbicides may damage soil microorganisms. There is disagreement in the literature on this issue. Sommers and Biederbeck (1973), after reviewing numerous laboratory and field studies, assert that herbicides (and insecticides), when applied at recommended field rates, generally have no lasting
harmful effects on microbial populations in the soil.

Greaves (1979) appears to be in at least partial agreement with Sommers and Biederbeck. Summarizing a number of studies of the effects of
herbicides on soil microorganisms, Greaves states that "while there is
some evidence that changes in the soil microflora do occur following long-
term use of herbicides, the data suggest that these changes do not necessarily result in significant changes in soil processes such as nitrogen
transformation. Where processes do apparently change, this is generally
ascribable to decreases in soil organic matter or to uptake of nutrients
by surviving weeds" (Greaves, 1979, p. 131).

Greaves does not take this conclusion as grounds for complacency, however. He points out that most of the work on the problem has ignored the
effects of herbicides on microorganisms living on roots. Herbicides affect
root morphology and may, therefore, damage these root dwelling microorganisms. Greaves ends by noting that relatively little is known about the
long-term effects of herbicides and calls for more research.

Eijsackers and van der Drift (1976, p. 169) make a similar argument.
They assert that, while field tests show that soil fauna recover quickly
from the toxic effects of herbicides, the processes underlying this are
not well understood and "extensive research" is needed. Von Rumker and
coauthors (1975) state that most studies of the metabolism and degradation

of herbicides have focused on effects of herbicides residues in the soil on following crops or other valuable vegetation. Consequently, there is little information about other effects of herbicides residues in the soil or in other elements of the environment.

There is enough uncertainty about the more subtle, long term effects of herbicides on soil and water dwelling organisms, and perhaps on humans, to warn against complacency about present environmental impacts of these materials. Still, the available evidence does not indicate that these impacts impose major environmental damage. This judgment is supported by a survey of thirty-six agricultural scientists from all around the country (Barnes and coauthors, n.d.). The scientists were asked to describe cases of water pollution by pesticides which in their judgment constituted a problem. Only two problems clearly involving farm applied herbicides were listed. Both were crop damage resulting from use of herbicide-contaminated water for irrigation. Picloram, a particularly persistent and leachable herbicide has been found in the water of wells in two counties in Nebraska. The concentration was only a few parts per billion, but enough to damage crops irrigated with the well water. Similar damage was reported in cases where water from a pond contaminated by herbicides in runoff was used for irrigation, and where herbicides-contaminated runoff flows directly into fields.

The survey also mentioned possible damage from atrazine and alachlor residues in sediment in the Chesapeake Bay estuary and in the Bay itself. A study suggested that these residues were killing bottom vegetation in the Bay. This is in dispute, however, because the scientific procedures used in the study are believed by some to be inadequate.

R. D. Wauchope, the person responsible for the survey of expert opinion about present pesticide damages to water quality, summarized his findings as follows:

> It is unlikely that any significant observable nonpoint source water quality problem due to proper agricultural or silvicultural use of pesticides would not be known to at least one of the sources [interviewed]...It is a reasonable conclusion, then, that with a few possible exceptions, currently-registered agricultural and silvicultural pesticides are not observed to be causing problems with respect to water quality (Barnes and coauthors, n.d., p. 15, emphasis in the original).

This conclusion must be taken with certain caveats. The experts surveyed by Wauchope did not address the possibilities that some herbicides, as noted above, may be carcinogens, mutagens or teratogens; and we already have seen that the absence of evidence that herbicides now constitute a major environmental threat may simply reflect the failure of researchers to look in the right places, or the inability of present detection techniques to pick up effects at very low trophic levels. Nevertheless, Wauchope's survey supports the judgment that, on the available evidence, present levels of use of herbicides do not pose major threats to the environment.

Future Severity of the Problems

The future environmental impacts of pesticides will depend upon the quantities of these materials used by farmers and upon their characteristics, particularly their toxicity and longer term effects on humans and other non-target organisms. We discuss both quantities likely to be used and characteristics.

Insecticides. Judging from present quantities and non-persistent characteristics of insecticides used, the environmental impacts of these materials are concentrated primarily in the Southeast and Delta and secondarily in the Corn Belt. This reflects the location of cotton production in the Southeast and Delta and the difficulty of controlling cotton insect pests in those regions, and the need to control corn insect pests, particularly the corn rootworm.

We concluded in the last chapter that the quantity of insecticides applied to cotton would decline from present levels and that applications to corn would increase little if at all. These crops are so dominant in total insecticide use that these results imply a significant reduction in the total quantity of insecticides used unless there were an extraordinary increase in amounts used on other crops. Our analysis of trends in insect management technologies for wheat pointed to no such increase. On the contrary, these trends are toward diminishing per acre applications of insecticides. While our projection of land in wheat indicates an offsetting trend, the prospect, in our judgment, is for little if any increase in total insecticide applications to wheat.

We projected an increase in insecticide use on soybeans because of a relative shift of soybean acreage to the Southeast and Mississippi Delta. The amount of insecticides applied to soybeans will remain small, however, and the projected use will be more than offset by the anticipated decline in insecticides used on cotton and corn.

The implication is that so far as the environmental problems of insecticides are a function of quantities used, the problems should diminish significantly, both from the national perspective and from the perspective of the three most affected regions.

We expect the characteristics of insecticides also to change in ways tending to diminish environmental damage. The shift away from the organochlorine compounds appears permanent, indicating that the particularly troublesome problems associated with persistence, biomagnification, and bioaccumulation will continue to decline.

The substitution of the organophosphorus and carbamate compounds for the organochlorines substitutes problems of acute toxicity for those of persistence, biomagnification, and bioaccumulation. To the extent that human mortality and morbidity are measures of environmental impact, this substitution would appear to be for the worse. While this seems a plausible presumption, such data as are available--and they are quite incomplete-- do not indicate an increase in insecticide related deaths associated with the shift to organophosphates and carbamates (National Academy of Sciences, 1975a, pp. 87-89). Indeed, as noted earlier, the number of accidental deaths from pesticide poisonings was less in 1974 than two decades previously, the period when the shift from the organochlorines to organophosphorus and carbamate compounds was occurring (Pimentel and coauthors, 1980).

The shift away from the organochlorines may also have reduced water pollution by insecticides. Pimental and coauthors cite three studies of pesticide concentrations in surface waters of the United States which show a steady decline in concentrations from 1964 to 1978. Total pesticide use increased sharply in this period. These authors attribute the decline in concentrations to the shift from the persistent organochlorines to the less persistent organophosphorus and carbamate compounds.

There is another, perhaps subtle, respect in which the shift to the organophosphorus and carbamate compounds may be favorable to the environment. Injury to humans and other nontarget organisms by these materials result from their acute toxicity, meaning that the injury is severe and quickly evident. Consequently, both the victims and those responsible usually can be readily identified. The damages caused by the organochlorine compounds, by contrast, do not produce sudden, severe symptoms but rather show up as impaired reproductive processes, cancer, or other long-delayed effects. In addition, because of the persistence, biomagnification and bioaccumulation properties of the organochlorines, it may be difficult if not impossible to determine the original source of the damaging materials and those responsible for them. Because of these differences in the timing and nature of the damages done, the organophosphorus and carbamate compounds likely would be easier to manage than the organochlorines. The point is not that the potential environmental damages of the organophosphorus and carbamate compounds are less--they may in fact be greater--but that the actual damages may be less because of greater ease in management of these materials.

The substitution of synthetic pyrethroids for some of the organophosphorus compounds in control of cotton insect pests seems likely to continue. Since the pyrethroids have low toxicity to mammals and are not persistent, this trend reduces pressure on the environment. Their apparent high toxicity to fish is offset by the low probability that they will reach surface waters in significant amounts. There are reports from the Southwest of emerging resistance of the tobacco budworm to the pyrethroids, and if this becomes general, the usefulness of these materials will be greatly reduced. If this happens, however, it would appear to be 5 to 10 years ahead, time in which to develop substitutes for the pyrethroids. While we do not now foresee what these substitutes may be, the whole thrust of research and development in cotton pest management technologies is toward substitution of nonchemical for chemical means of control. We consider it unlikely, therefore, that whatever substitutes for pyrethroids are developed--if indeed they become necessary--will pose a greater threat to the environment than the pyrethroids.

We conclude that on balance changes in characteristics of insecticides will reinforce the prospective decline in amounts applied to significantly reduce the environmental damages of insecticides over the next several decades.

Herbicides. We concluded in the previous chapter that if per acre amounts of herbicides applied to land in corn, wheat, soybeans, and cotton remain the same as in 1976, total herbicide applications to these crops (82 percent of all herbicides applied to crops in 1976) would rise to 53 percent from 1976 to 2010 (see table 6-13). We also noted that if conservation tillage spreads to 50 to 60 percent of cropland by 2010, as seems likely, the increase in total herbicide use would be significantly higher than these numbers indicate.

While the shift to conservation tillage will increase the amounts of herbicides applied more than otherwise would occur, the movement of herbicides from farmers' fields probably will not rise in proportion to the increase in amount of herbicides. The reason is that losses of herbicides through erosion and runoff probably are less with conservation tillage than with conventional tillage. This is the conclusion of a survey of the literature on tillage and losses of pesticides by Wauchope, McDowell and Hagen (n.d.). It was the key finding also of a study of small watersheds in Iowa by Baker and Johnson (1977).

Despite the apparent favorable effect of the shift to conservation tillage in reducing herbicide losses, the projected increase in amounts applied in 2010 is so large that a significant increase in the quantity of herbicides in the environment is implied. To the extent that quantity is a measure of environmental burden, that burden would increase.

The issue of the inherent threat of herbicides to the environment thus is crucial in judging the severity of the future threat. We concluded from our previous discussion of the issue that the evidence does not support the inference of a major inherent threat. We noted, however, that some aspects of the herbicide-environment relationship which may contain a threat have not been thoroughly investigated. The possibility of real or potential threats, therefore, cannot be discounted. Moreover, we have no experience with the use of herbicides on the scale projected for 2010. Conceivably, there could be thresholds of environmental damage which expansion on the projected scale would pass.

The possibilities of presently undetected or potential future environmental damages of herbicides are sufficiently likely, in our judgment, to justify intensive, continuing investigation of herbicide-environment relationships, and a wary attitude toward the expanding use of herbicides. Adopting such an attitude, we nevertheless conclude that present knowledge does not suggest that the projected expansion of herbicide use will pose a major threat to the environment.

Irrigation

Nature of the Environmental Impacts

Irrigation may, and in arid areas typically will, lead to an increase in soil salinity. Where water is used several times for irrigation, as river water usually is, the salt content of the water also rises. Too much salt in the soil or in irrigation water inhibits plant growth, and in extreme cases may render the land and water useless for agriculture.

All water used for irrigation carries salts. The increasing concentration of salt in the soil occurs because of evaporation of some of the water applied to the land. Over time the amount of the salt will increase unless it is periodically flushed out by rain or by water applied specifically for that purpose. Flushing of the salt will protect the productivity of the land so cleansed, but it increases the salt content of the water available to farmers downstream. The salt content of the water also increases as it moves downstream because irrigation return flows carry salt from the land to the river.

If drainage is inadequate, repeated irrigation may also raise the underground water table within reach of the plant root zone. Capillary action then will carry water close to the soil surface where it evaporates, leaving a salt residue. The accumulation of these residues eventually will reduce the productivity of the land.

Streamflow is reduced by withdrawals of river water for irrigation. The decline in streamflow increases pollutant concentrations and water temperature. Both can have devastating effects on riverine flora and fauna. Moreover, the flow of fresh water into estuaries and tidal marshes is diminished, reducing the productivity of these ecologically vital and often commercially valuable resources.

Irrigation may increase erosion relative to pre-irrigation treatment of the land. It may also have the opposite effect, however, by increasing the plant cover of the land, thus reducing its exposure to the erosive effects of wind and rain.

Because irrigation increases the productivity of the land compared with dryland farming, it justifies greater per acre use of fertilizer and pesticides, thus increasing the likelihood of environmental damage from these materials.

Present Severity of the Problems

Salinity is the most pervasive environmental problem stemming from irrigation in the United States.[12] Salt levels are high and generally rising in all western river basins except the Columbia. In much of the Lower Colorado, parts of the Rio Grande, and the western portion of the San Joaquin river basins, the salt concentrations in either the water or the soils are approaching levels that threaten the viability of traditional forms of irrigated agriculture. Although groundwater salinity is not now a widespread problem in the West, the problem is growing and already is serious in parts of California, New Mexico, Montana, and Texas (U.S. Department of Interior, April 1975, pp. 116-118; and U.S. Water Resources Council, 1978, vol. I, summary, p. 65).

van Schilfgaarde estimates that roughly 25 to 35 percent of the irrigated lands in the West have some type of salinity problem, and that the problems are getting worse.[13] The problem is particularly severe in irrigated areas of the Lower Colorado River Basin and the west side of the San Joaquin Valley in California. The underlying cause of the problem in the two areas differs. In the Colorado, about one-third of the salt delivered in the lower reaches is from irrigation return flows, the other two-thirds coming from saline formations the river passes over. Elimination of deliveries from these natural sources would require expensive investments to divert the river and its tributaries around some of the areas contributing the salts. The salts from irrigation could be curtailed greatly by reducing irrigation return flows. Annual damages from

---

[12]This discussion of salinity in the West is taken from Frederick (1982).

[13]Jan van Schilfgaarde, director, U.S. Salinity Laboratory, Riverside, California in an interview with Kenneth Frederick, February 1980.

salinity in the Colorado River were estimated between $75 and $104 million in 1980 (U.S. Department of Interior and U.S. Department of Agriculture, 1977).

In the San Joaquin Valley, the principal salinity problem occurs because poor drainage prevents salt-laden waters from being carried away from the fields. High water tables already threaten the productivity of about 400,000 acres, and ultimately more than 1 million acres in the valley may be similarly affected. A $1.26 billion drainage system has been proposed to carry irrigation runoff from the western side of the valley to the Sacramento Delta. Because farmers would have to install their own underground drainage to get the waters to the central drain, the total costs of an effective system would be considerably higher than $1.26 billion (U.S. Department of Interior, California Department of Water Resources and California State Water Resources Control Board, 1979).

Salinity induced by irrigation is overwhelmingly a problem of the arid and semiarid west. It is not a significant problem in the principal irrigated areas of the east: the Mississippi Delta region, the Great Lakes states and the Southeast.

The contribution of irrigation to the nation's erosion problem is minor. Although erosion and sedimentation are viewed as problems in many areas of the west, it is not clear, with a few exceptions, that irrigation is an important contributor to these problems. A report by the U.S. Department of Interior (April 1975, p. 127) states that in the eleven most western states "most of the rapidly eroding range, grassland, and forest-covered soils occur where natural geologic erosion is dominant." The human contribution to erosion probably has been more important in the Plains States, but the contribution there of irrigation is mixed. Center pivot irrigation is important in that region, and the wheel tracks left with this system can lead to gully erosion. However, erosion by wind is far more important than erosion by water in the west, and irrigation, by increasing ground cover, helps to contain wind erosion. The National Resources Inventory showed that soil loss from wind erosion on cropland in the 10 Great Plains states was 893 million tons in 1977, or 5.3 tons per acre (USDA, August 21, 1979; and February, 1980). Erosion by water (sheet and rill erosion) in these states was 516.6 million tons, or 3.1

tons per acre. Rain, not irrigation, is the dominant cause of water erosion in those states.

The Soil Conservation Service sets 3 to 5 tons per acre per year as the maximum amount of soil loss consistent with maintaining the productivity of the land indefinitely. By this standard, we can conclude that irrigation in the west does not contribute importantly to the region's erosion problem. Indeed, the NRI data suggest that erosion of cropland by water, whatever the source, is not generally a problem in the west.

We point out below that sheet and rill erosion is a major problem in parts of the East, including parts of the Mississippi Delta and the Southeast, the principal areas of irrigated farming in the east. However, in those Delta and Southeastern states where erosion is particularly serious--Mississippi, Alabama, and Georgia--less than 10 percent of cropland is irrigated, and there is no reason to believe that irrigated land contributes more erosion per acre than nonirrigated land. Indeed, except when sprinkler systems are used, irrigation typically requires that the land be leveled, suggesting that irrigation would reduce erosion from such land.

The second National Water Assessment (U.S. Water Resources Council, 1978) identifies agricultural chemicals as a "significant" source of surface water pollution in three areas in the west, but does not indicate whether irrigated or dryland farming is the principal culprit. In comparison with the rest of the country this problem is not very extensive in the west. In our earlier discussion of fertilizer pollution we noted certain "hot spots" in southern California where nitrates in groundwater are perceived to be a problem, with irrigation a probable contributor. We also noted nitrate levels of 20 to 25 ppm in groundwater in the Sandhills of Nebraska, where irrigation is extensively practiced. However, in other parts of the Plains states that have been under irrigation for many years, e.g., the High Plains of Texas, there still are no observable problems of groundwater pollution from irrigation.

The Assessment indicates that agricultural chemicals present more water quality problems in the east than in the west. However, since only a small percentage of cropland in the east is irrigated, the contribution of irrigation to these problems, such as they are, must be correspondingly small.

Future Severity of the Problems

We expect only limited further expansion of irrigation in the west, so we do not expect the addition of irrigated land to significantly increase the environmental damages of irrigation. However, the severity of the major source of damage, salinity, is not just a function of the amount of irrigated land. Salinity tends to increase on a given amount of land with repeated irrigations. The salinity problem, therefore, is likely to increase more than in proportion to the increase in irrigated land unless steps are taken to deal with it. According to Frederick, improved farm management techniques can go a long way toward reducing some salinity problems and obviating the need for some of the costly structural solutions under consideration.[14] Efficient water and agronomic management reduces evaporation losses, permitting achievement of given yields with less water. The reduction in water applied and in evaporation slows the buildup of salt concentrations in the soil. In addition, agricultural scientists have developed crops and irrigation techniques which enable farmers to irrigate successfully with surprisingly high salt levels (e.g., van Schilfgaarde, 1977).

Improved basin-wide water management also has great potential for mitigating salinity problems. Where poor drainage prevents salt-laden waters from being carried away from the fields, artificial drainage is required to prevent eventual loss of productivity. But the extent of the drainage needs can be greatly reduced through good management. van Schilfgaarde suggests that the San Joaquin Valley drainage problem could be reduced to about 5 percent of current levels and a long-term equilibrium reached through an integrated irrigation system in which the best water is used first on salt-sensitive crops with the increasingly salt-laden runoff applied to increasingly salt-tolerant crops. The remaining highly saline waters would be reduced to quantities that could be disposed of in evaporation ponds rather than requiring a costly trans-basin drain.[15]

---

[14]This discussion of techniques for managing salinity is based on Frederick (1982).

[15]van Schilfgaarde in an interview with Kenneth Frederick, February, 1980.

We concluded above that irrigation does not contribute importantly to erosion at present, and we do not expect this to change over the next several decades. Part of the considerable increase in fertilizer applications we project for the Southeast, however, is associated with the prospective expansion of irrigation in that region (see chapter 3), much of which will be on sandy soils in Georgia and Florida where percolation rates are high. The expansion of irrigation in that region, therefore, likely will contribute to increasing nitrate concentrations in groundwaters of the region.

Erosion

Nature of the Environmental Impacts

There are two sorts of impacts of erosion: on-farm and off-farm. Off-farm impacts of erosion by water include the sedimentation of reservoirs and lakes, rivers, and harbors, and the need to process the water to make it drinkable or usable for industrial purposes. Suspended sediment also reduces recreational values provided by water and may injure fish populations. When sediment settles, it shortens the useful life of reservoirs for generation of power and in providing irrigation and flood control, and of harbors in all the manifold services they provide. Removal of the sediment from these water bodies becomes a continuing and costly burden typically borne by taxpayers. Soil carried by wind also imposes off-farm costs of clean-up, damages esthetic values and, where particularly thick, may constitute a traffic hazard.

The off-farm costs of erosion are examples of "externalities"; that is they are not borne by the farmer whose eroded soil causes them but by other members of the society. Since the farmer escapes these costs he has no incentive to control erosion to reduce them. If those bearing the costs are unable through private action to induce the farmer to take corrective measures, there is a clear case for public intervention to accomplish this. The main issue then is whether the costs of reducing erosion are less than the costs of the damages.

The on-farm cost of erosion is the loss in productivity resulting from the loss of soil. The soil supports plant growth by providing storage for nutrients, moisture and air, and a medium in which plant roots can take hold. Erosion will eventually reduce the capacity of the soil to provide these services to the plant unless new soil is formed at a rate sufficient to offset erosion losses. The rate at which new soil is formed is highly variable, depending upon the nature of the soil parent material, climate, biological factors (plants, soil flora and fauna) and topography (Cady, 1980). The U.S. Soil Conservation Service (SCS) has defined the tolerable level of soil loss ("T" value) as the maximum amount of loss per acre per year "that will permit a high level of crop productivity to be sustained economically and indefinitely" (Wischmeier and Smith, 1965, p. 2). Both physical and economic factors are considered in establishing "T" values, although Wischmeier and Smith's discussion suggests that physical factors are dominant. They assert, for example, that "T" values in the United States range from 1 to 5 tons per acre per year, depending upon soil properties, depth, prior erosion, and topography. Five tons of topsoil is about one-thirtieth of an acre inch, suggesting that in the judgment of the SCS that is the maximum amount of new soil that can be formed in a year in most circumstances.

It is clear that sustained erosion in excess of the rate of soil regeneration will eventually reduce the productivity of the land, given technology, management, and other factors affecting productivity. The SCS to the contrary not withstanding, however, it is not clear that soil loss in excess of soil regeneration justifies action, immediately or ever, to achieve a balance between loss and regeneration. Reducing erosion is not costless. To achieve it the farmer must put in terraces, or windbreaks, take fragile land out of production, or adopt other measures which either reduce his production or increase his costs. Public investment in research to develop land-saving, erosion-reducing technologies is another alternative, obviously not costless, for reducing erosion.

Judgments about when, where, and how much erosion control is justified to protect productivity clearly require comparison of costs of lost productivity with costs of erosion control. For many farmers, and certainly for soil conservationists acting to protect society's interest in the land, these judgments are tempered by a conservation ethic, the stronger the ethic the greater the willingness to bear control costs. But farmers inevitably must be sensitive to the balance of costs and benefits of erosion control, and conservationists cannot responsibly be indifferent to them.

Thus farmers can be expected to adopt erosion control practices when the balance of costs and benefits favors this, and for many dedication to the conservation ethic likely gives added incentive. Yet there is a widely held presumption that left to their own devices farmers will not adopt on a scale consistent with the public interest in maintaining the productivity of the land. Why might they fall short? There are two general reasons: (1) the present value of the land in agricultural production as reflected by the market may underestimate its present value to society. In this case the farmer will invest less in maintaining the productivity of the land, or in restoring the productivity of eroded land, than is socially desirable.[16] (2) The market may accurately measure the social value of the land but, for a variety of reasons, farmers may not adopt conservation practices on the socially desirable scale.

The notion that the market may undervalue agricultural land runs contrary to the commonly held view that at least for the last decade or more farmland in the United States has been overvalued. All around the country net returns to the land from farming are low relative to its market price.

---

[16] In commenting on this study, Glenn Johnson put great emphasis on investments to restore the productivity of eroded land as a key element in judging the social importance of erosion. Johnson stresses that the land's capacity to produce depends both on the productive properties of the soil and on the capital, including human capital, joined with it. Over a period of years the tilth, organic matter, nutrients, and other productive properties of even badly eroded soils often can be restored, e.g., by putting the land in a periodically plowed-under cover crop and applying lime, fertilizer and manure.

As a response to erosion, investments to restore the land's productivity clearly are an important adjunct to measures to reduce erosion. While data are lacking, there can be little question that in the United States such investments have contributed to agricultural productivity. From the standpoint of policy, the important question about these investments, as about measures to reduce erosion, is whether farmers, responding to market signals, will undertake them on a scale consistent with the social interest in the productivity of the land.

The reason is that the land has value apart from its contribution to agricultural production. Around urban areas agricultural land prices are bid up to reflect the value of the land in non-agricultural uses. More widely, high inflation relative to interest rates in much of the 1970s gave farmers and others incentive to acquire debt, and land was a prime medium for doing this. Castle and Hoch (1982) have shown that this was a major determinant of the price of agricultural land in the 1970s.

The relatively high price of land in the last decade thus does not indicate that the market overvalues its future contribution to agricultural production. So, the high price gives no special incentive to farmers to protect the productivity of the land for that purpose. On the contrary, farmers around urban areas who expect to eventually sell their land to developers may be indifferent to the productivity effects of erosion since these are unlikely to significantly affect the ultimate selling price. Similarly, farmers whose main interest in buying land is to acquire debt likely will not see the high price as a signal to undertake erosion control measures.

For our purposes here the question is whether the market determined net returns to the land from agricultural production understate the social value of the land in that use. If the answer is yes then farmers' incentives to adopt erosion control practices will be weaker than is consistent with the public interest in the productivity of the land.

Why the market may undervalue the land. There are five main reasons: (1) a general lack of knowledge of the effects of erosion on future yields resulting in a systematic underestimate of the effects and a corresponding overestimate of the future supply of land; (2) the market misjudges the strength of forces affecting the future demand for food and fiber and so underestimates future prices; (3) the market overestimates the rate of emergence of economical land-saving technologies; (4) the social cost of investments in erosion control measures is less than the private costs; (5) the market gives less weight than society to maintenance of the productivity of the land as a hedge against future demand for food and fiber.

The first three sources of error arise because of ignorance among those in the society who make the market for agricultural land. To make the case for public action it is necessary to show that those acting in society's interest are less ignorant than those who make the market about the effects of erosion on productivity, future prices of food and fiber, and the rate of emergence of economical land-saving technologies. We are not persuaded that those acting for society are smarter than the market with respect to any of these three factors. Studies of the effects of erosion on yields under the great diversity of soil and climatic conditions under which American farmers actually operate are relatively few so policy makers have little to go on. Each farmer, however—at least the sizable percentage who are owner-operators—has long experience and detailed knowledge of his particular situation, and most farmers will be well informed also about their neighbors' experience. Farmers moreover have a direct economic interest in informing themselves about the yield effects of erosion. It would be surprising if under these circumstances non-farmers, no matter how technically competent, were more knowledgeable than farmers about the effects of erosion on yields.

The other two factors concern future events—prices of food and fiber and technological change—about which everyone is ignorant. The record does not demonstrate that those acting for society are more skillful than the market in forecasting these events.

If the rate of interest appropriate for public investments is less than that for private investments then the social cost of investment in erosion control will be less than the private cost, an argument for public intervention. There is no need to review here the large literature on differences between public and private rates of discount. It is enough to note that there is an argument, although a weak one in our judgment, for using a lower rate in calculating costs of and returns to public investments. The key element in the argument is that the taxing power of government means that government debt is not secured only by performance of public investment projects. Consequently the risk of default on government debt is less than on private debt, so governments can borrow more cheaply. The point is not that for a given project the risk of failed performance is less if it is publicly financed than if it is privately financed. The difference is in the ability to repay in the event of failure. This being the case, does the lower rate indicate that social costs are less than private costs? We are skeptical that it does.

The fifth reason why the market may understate the social value of the land in agriculture is that society may want to give more weight than the market to maintenance of the productivity of the land as a hedge against future demands for food and fiber. The point here is not simply that the future is uncertain. Both those who make land markets and those acting for society's interest in the land know that future demand-supply conditions affecting the land are uncertain. Both may agree perfectly on the bounds within which these conditions will fall, e.g., with 95 percent probability. But the market price of the land will represent a consensus view of these conditions while those acting for society may feel obliged to respond to conditions indicating more pressure on the land than reflected in the consensus view. With respect to future demands on the land, for example, society may want to give more weight than the market to possible high demand outcomes. Within the range of generally agreed rates of land-saving technological change, those acting for society may feel constrained to give more weight to low rates of change.

Among the various arguments why the market may understate the social value of the land in agriculture this last strikes us as most persuasive. It does not assume that those acting for society's interest in the land know more than anyone else about future conditions affecting the value of the land. Rather the basis of the argument is that the responsibilities of those acting for society require them to give more weight than the market to the probabilities of unfavorable outcomes with respect to the growth of demand for food and fiber or the emergence of economical land-saving technologies.

The argument does not rest on the notion that society has a longer time horizon than the farmer in land management decisions. The difference between society's interest and the farmer's interest in soil conservation practices occurs because society assigns higher value to the annual net benefits of these practices, not because the period over which the benefits are discounted is different. In calculating the effect of the benefits on the price of land both society and the market discount them into perpetuity. If the market's perception of the value of the land comes into line with society's perception, the value of the practices will rise and the farmer will have incentive to adopt them, whatever his time horizon.

Why farmers may not react. The statements just made about farmer behavior assume that farmers are economically rational in the usual sense; that if the effects of erosion and of the costs and benefits of erosion control are reflected in the value of their land, they will know it; that they bear the long-term costs of not controlling erosion as well as the costs and benefits of control; and that they are not constrained by lack of capital from adopting erosion control practices. If these conditions are not met farmers are not likely to undertake erosion control practices on the socially desirable scale even if the price of land reflects the full social value of the practices.

We are comfortable with the assumption of economic rationality. This of course does not mean that farmer behavior is shaped only by economic criteria. It does mean that in making farm management decisions farmers give heavy weight to these criteria. The belief that farmers do not adopt erosion control measures because they are economically irrational is a bad guide to policy.

We also are comfortable with the assumption that if the market price of the land captures the effects of erosion and of the benefits and costs of erosion control farmers will know it. Land is easily the most important single asset of most farmers. They are certain to be sensitive to changes in its value and to seek information about the factors affecting value, particularly those which the farmer himself can influence if not control. If farmers fail to respond to erosion and erosion control measures, the reason is far more likely to be because these are not reflected in the price of the land rather than because of farmer ignorance of their effect on price.

Tenants are not likely to bear the long-term costs of erosion nor to capture all the benefits should they invest in erosion control. To the extent that tenants rather than owners make land management decisions, investment in erosion control is likely to be less than it otherwise would be. But if in fact erosion control is justified why would owners not require that tenants adopt control practices as a condition of tenancy, or invest in these practices directly? Most absentee owners have other, perhaps non-farm, interests and may conclude that achieving more erosion control in the farm would be more costly of their time and other resources than is justified. But in this case, is the amount of erosion control less than socially desirable? It may be that if tenancy were abolished erosion would be reduced. But the institution of tenancy presumably serves some useful social purpose, so abolishing it would impose some social costs. If these are higher than the costs of erosion then abolishing tenancy would be a poor way to achieve erosion control.

Farmers may under-invest in erosion control measures because of capital constraints. In an economically perfect world capital will be available for all investment projects which earn at least the socially optimal rate of return. The world is not perfect, however, economically or otherwise, and some farmers may be denied the capital necessary for privately and socially optimal investment in erosion control. Where this occurs a case can be made for public intervention to remove the capital constraint.

In summary, there are some arguments for believing that farmers may invest less than socially desirable amounts in control of erosion to protect productivity. The strongest of the arguments, in our judgment, is that society has a responsibility for stewardship in management of agricultural land while the market has none. The implication is that society should place a higher present value than the market on agricultural land for agricultural production.

We believe the argument has weight. However, as a guide to soil conservation policy it has an important limitation. It establishes a presumptive case for public action but provides no guidelines as to the scale of the effort. The argument asserts that market determined rates of soil loss are excessive by a social standard but it does not define the standard. It is not necessarily that set by the SCS--soil loss no greater than soil regeneration--since the social cost of achieving the SCS standard may be more or less than the social benefits.

The argument for erosion control policies to protect productivity thus leaves much room for judgment in specifying socially acceptable standards of soil loss. The judgments differ not only because people have different expectations about the future but also because some give more weight to the conservation ethic than others. Since the future is inherently uncertain and ethical standards not easily changed by argument, the range of judgments is likely to remain wide and a consensus about the need for and effectiveness of conservation policies elusive. We return to this set of issues in our discussion of policies in the next chapter.

Present Severity of the Problems

There are no reliable, comprehensive estimates of past or current erosion induced losses of productivity, much less estimates of whether the losses were, or are, excessive from the social standpoint. Whatever past losses may have been, advances in technology and investments in land restoration more than compensated for them.[17] Crop production per acre of cropland increased 110 percent from 1910-14 to 1975-79, an annual rate of 1.14 percent. The rate of increase was especially rapid after World War II: 1.83 percent.

The increase in yields does not demonstrate that productivity losses from erosion were negligible from a social standpoint. Many conservationists insist that they were not. They argue, in effect, that the social interest would have been better served if in expanding output farmers had depended more on preserving productivity by reducing erosion and less on finding technological substitutes for the land. For reasons already given, the argument rests on judgments not easily evaluated. The fact is no one knows whether productivity losses to erosion were socially excessive over the last several decades.

Reliable estimates of off-farm costs of erosion from agricultural land also are in short supply. Pimentel and coauthors (1976) cite various sources indicating that costs of dredging rivers and harbors, of reduced useful life of reservoirs, and of other sediment damages came to about $500 million annually in the 1960s. In 1980 prices, these costs probably would not be less than $1 billion.

However, these costs reflect total erosion, not just that from agricultural land. Pimentel and coauthors estimated total water borne sediment at 4 billion tons per year, three-quarters of it from agricultural land. Meister and coauthors (1976) use figures which imply substantially smaller amounts of water borne sediment from agricultural land than estimated by Pimentel and coauthors. In contrast, estimates from the NRI indicate that the Pimentel figures may be low.

---

[17]See footnote 16.

Clearly there is great uncertainty about present on-farm and off-farm costs of erosion. Nevertheless, consideration of current rates of erosion provide some insights to the present severity of the problem. Since in this study we are interested in the environmental impacts of crop production, we focus on erosion from cropland.

Table 7-1 shows national and regional estimates of erosion from cropland in 1977. For the nation, total erosion was 2,799 million tons, about two-thirds of it sheet and rill erosion and one-third erosion by wind. Sheet and rill erosion accounted for a little more than two-thirds of the 6.8 tons per acre of total erosion. Sheet and rill erosion also was important on rangeland, pasture, and grazed forest land (figures not shown in table 7-1), but per acre loss rates were significantly lower in those cases. Moreover, the heaviest losses occurred in less populated areas and on less valuable land, making their economic costs and damages to the human environment clearly less significant. Cropland erosion losses, by contrast, were concentrated in such productive states as Iowa, Illinois and Missouri (parts of the Corn Belt); Nebraska and Kansas (parts of the Northern Plains); Tennessee and Kentucky (parts of Appalachia); and Mississippi and Alabama (parts of the Mississippi Delta and the Southeast). Thus, the agricultural heartland of the country is most severely affected.

In the Great Plains, cropland also is subject to significant wind erosion. Texas, New Mexico, and Colorado are most vulnerable, with losses amounting to 9 to 15 tons per acre in those states. Indeed, wind erosion from cropland in Texas accounts for 16 percent of total cropland erosion in the entire nation. In Plains states other than Texas, New Mexico and Colorado, wind erosion rates fall below 5 tons per acre for all cropland.

The trend in erosion loss is difficult to assess because there are no reliable estimates of erosion prior to those in the 1977 National Resources Inventory (USDA, February 1980a). However, comparison of the 1977 NRI with a similar Conservation Needs Inventory taken by the USDA in 1967 shows that the amount of cropland needing conservation treatment fell by 3 percentage points between the two years--a period when the reserve of idle and often erosion-prone acreage was brought back into cultivation and many decried the apparent increase in erosion (USDA, 1974; USDA, February 1980a). This reported improvement in the face of greater cropping pressure

Table 7-1. Erosion from Cropland in the United States

Region	Wind Amount (mill.tons)	Wind Tons per acre	Sheet and Rill Amount (mill.tons)	Sheet and Rill Tons per acre	Total Amount (mill.tons)	Total Tons per acre	Percentage of total Erosion	Percentage of total Cropland
Nation	891	2.1	1,908	4.7	2,799	6.8	100	100
Northeast	n.e.		82.9	5.0	82.9	5.0	3.0	4.0
Lake States	n.e.		117.5	2.7	117.5	2.7	4.2	10.7
Corn Belt	n.e.		688.3	7.7	688.3	7.7	24.6	21.8
Iowa	n.e.		261.3	9.9	261.3	9.9	9.3	6.4
Northern Plains	212.3	2.2	322.4	3.4	534.7	5.6	19.1	22.9
Nebraska	25.9	1.3	117.8	5.7	143.7	7.0	5.1	5.0
Appalachia	n.e.		186.3	9.0	186.3	9.0	6.7	5.0
Tennessee	n.e.		69.5	14.1	69.5	14.1	2.5	1.2
Southeast	n.e.		111.0	6.3	111.0	6.3	4.0	4.2
Georgia	n.e.		42.7	6.6	42.7	6.6	1.5	1.6
Delta	n.e.		154.9	7.3	154.9	7.3	5.5	5.1
Arkansas	n.e.		46.7	5.9	46.7	5.9	1.7	1.9
Southern Plains	488.8	11.6	141.4	3.4	630.2	15.0	22.5	10.2
Texas	453.5	14.9	99.5	3.3	553.0	18.2	19.6	7.4
Mountain	190.3	4.5	70.8	1.7	261.1	6.2	9.3	10.2
Pacific	n.e.		31.9	1.4	31.9	1.4	1.1	5.6
California	n.e.		8.6	.9	8.6	.9	.3	2.4

Note: n.e. means not estimated

Source: U.S. Department of Agriculture, Basic Statistics, 1977 National Resources Inventory (Washington, D.C., Soil Conservation Service, 1980a); and "RCA Appraisal 1980 Review Draft," Pt. I and II (Washington, D.C., USDA, 1980b). The data are for the 48 contiguous states. When Hawaii and Caribbean area are included, total sheet and rill erosion is 1,926 million tons.

Sheet and rill erosion are caused by water. Sheet erosion is the movement of continuous layers of soil from the field. If the water moves fast enough it scours the soil and cuts small channels in the soil surface. The soil moved in this fashion is rill erosion.

coincided with a significant spread in the use of conservation tillage during the same period.

At the national level, sheet and rill erosion losses from cropland approach the maximum of 5 tons per acre per year set by the SCS as consistent with long-run maintenance of the productivity of the land, and erosion by wind and water combined exceed the limit.

These national soil loss averages conceal more severe local and regional problems. Many states, especially in the West and Lake states, have low average losses from cropland, while in parts of the Corn Belt, Delta, Southeast, and Southern Plains losses far exceed tolerable amounts. But erosion is a still more localized problem. It may be severe in very small areas or on individual fields or parts of fields, even in regions that generally are not afflicted. Nationally, 10 percent of the cropland accounts for 90 percent of all sheet and rill erosion in excess of 5 tons per acre (USDA, 1980b, Part II). Because the problem tends to be localized, it is amenable to sharply focused programs of control, discussed in the next chapter.

Erosion losses from the top layers of soil usually are the most damaging to soil quality because they are richest in the nutrients and organic matter that are most accessible to plants. Damage on the field can be measured in terms of the costs of replacing the nutrients lost and the possible decrease in the productive potential of the land once erosion exceeds "T" value. There are no confident measures of either effect. The RCA study concluded that erosion at 1977 rates would reduce potential corn and soybean yields in the Corn Belt by 15 to 30 percent over a fifty year period (USDA, 1980b, Part II). Other studies suggest that these figures may be conservative. This does not necessarily mean that yields will decline, but the land will become less responsive to other inputs and will require more of them to sustain any given level of output.

The RCA estimates suggest that indefinite continuation of present rates of erosion will in time impair the productivity of the land enough to increase costs of agricultural production. Since trends in prices and productivities of nonland inputs also suggest rising costs, the additional impetus to costs given by erosion is not to be taken lightly. In addition, as we already have seen, estimates by Pimentel and coauthors (1976) indi-

cate that off-farm erosion damages to lakes, reservoirs, and harbors probably are not less than $1 billion per year in 1980 dollars. There is bas-
is for believing, therefore, that the on-farm and off-farm environmental
impacts of present rates of erosion are cause for concern. Given our projections of production, land use and technology, what can we say about
future impacts?

Future Severity of the Problems

To obtain insights to future environmental impacts of erosion we engaged the help of the Center for Agricultural and Rural Development at
Iowa State University. Specifically, we asked the center to run our projections to 2010 of land in feedgrains, wheat, soybeans, and cotton
through the ISU model of the U.S. agricultural economy to obtain projections of erosion.[18] The results are shown in table 7-2. A description
of how the results were obtained will be useful for interpreting them.
We provided ISU with an initial set of projections to 2010 of harvested
land in feedgrains, wheat, soybeans, cotton, hay, and fallow, for the
nation and the 10 USDA producing regions. The projections were fed into
the model and the result was the set of projections of erosion called
run 1 in table 7-2. (The model yields many other outputs, but our interest was in erosion.) As indicated in the appendix to chapter 5, the model
is structured to give results by the 19 river basins shown in the table
rather than by the 10 USDA producing regions. Moreover, the model had
difficulty in accommodating the RFF projections of harvested land precisely as we had prepared them, so the run 1 results reflect a slightly
different cropping pattern than the one we had projected. There was no
difference in the land totals, however.

The ISU model allocates land and production by three kinds of tillage
technology and four kinds of conservation practices. The tillage tech-

---

[18]As noted in footnote 3 in the appendix to chapter 5, the ISU model
is a cost minimizing linear programming type. It permits analysis of a
variety of resource and environmental issues touching U.S. agriculture
from both regional and national perspectives. The model was used extensively by the USDA in its RCA work to explore the implications of various
scenarios about future levels of agricultural production and policies for
controlling erosion.

Table 7-2. Projections of Sheet and Rill Erosion from Cropland in
the United States in 2010

	Run 1		Run 2	
	Total	Per acre	Total	Per acre
Region	(mill. tons)	(tons)	(mill. tons)	(tons)
Nation	2,320	5.2[a]	3,537	8.3[a]
New England	2.6	3.3	1.5	2.2
Mid-Atlantic	123.9	10.6	121.5	10.7
South Atlantic-Gulf	465.6	13.5	435.9	16.9
Great Lakes	50.5	2.0	83.2	3.4
Ohio	152.2	4.0	357.6	9.4
Tennessee	48.1	8.5	52.6	15.2
Upper Mississippi	284.7	4.1	617.8	9.0
Lower Mississippi	217.1	8.7	398.3	17.8
Souris-Red-Rainy	52.1	2.7	48.0	2.6
Missouri	257.9	2.6	559.8	5.5
Arkansas-White-Red	287.9	5.5	408.8	7.8
Texas Gulf	262.8	8.6	336.7	11.8
Rio Grande	20.0	5.6	13.7	5.2
Upper Colorado	2.0	1.3	2.5	1.6
Lower Colorado	.8	.6	.3	.5
Great Basin	2.4	2.0	6.4	4.2
Columbia-North Pacific	85.2	4.9	88.2	5.2
California-South Pacific	4.4	.6	4.3	.6

Source: Runs of the Iowa State University model of the U.S. agricultural economy done for this study. The regions are the river basins draining the 48 contiguous states.

[a]The estimates of erosion per acre exclude 48 million acres of cropland not treated in the ISU model. When this land is included, as it
should be to make these estimates comparable to those in the NRI, per
acre erosion from cropland is 4.7 tons in Run 1 and 7.4 tons in Run 2.

nologies are (1) fall plowing with the moldboard plow, residue removed
(here called conventional tillage); (2) spring plowing, with or without
the moldboard, with residue left until spring (here called low conservation tillage); (3) spring plowing with something other than the moldboard,
with residue left year around (here called high conservation tillage).
The four conservation practices are straight row planting, contour farming, stripcropping and terraces, each of which may be practiced with any
of the tillage technologies.

Given projections of production, the model chooses among tillage technologies and conservation practices so as to minimize production costs. In run 1 the model allocated 17.2 percent of harvested land to conventional tillage and 82.8 percent to the two forms of conservation tillage, 34.9 percent to the "low" form and 47.9 percent to the "high" form.

Our analysis of tillage technologies led us to conclude that, by 2010, some 50 to 60 percent of cropland would be in some form of conservation tillage. Thus the model's projection of 83 percent appeared high to us. This view was reinforced by the knowledge that the model takes account of the long run effects of erosion on crop yields, a characteristic favoring conservation tillage over conventional tillage in scenarios projected several decades into the future. We think it probable that farmers give less weight to the long-term effects of erosion than the model does. We were particularly skeptical of the high percentages put in conservation tillage by run 1 of the model in the Ohio basin (86 percent) and the Upper Mississippi basin (98 percent). Our analysis of tillage technologies suggested that in much of the Ohio basin soils are too moist to support such a high percentage of land in conservation tillage and in much of the Upper Mississippi region (which includes most of Minnesota and Wisconsin in addition to most of Iowa and Illinois) too cool soil temperatures in the spring would be similarly limiting.

For these reasons we asked ISU to make a second run of the model, using the same projections of production and land use as in run 1, but with two changes in other respects: (1) let the model allocate production by region, and (2) impose upper limits on the amount of land that could be in high conservation tillage of 60 percent in the Ohio basin and of 70 percent in the Upper Mississippi. In all other regions the model was free to allocate land among tillage technologies in accordance with its cost minimizing principle.

As table 7-2 shows, the second run of the model projected substantially more erosion, both in total and per acre, than the first run. This occurred even though the total amount of land in crops was less than in run 1 by 17 million acres (4 percent) and the amount of land in some form of conservation tillage was the same as in run 1, 83 percent. The differ-

ence was that the amount of land in high conservation tillage fell from
48 percent in run 1 to 31 percent in run 2. In the Ohio basin land in
high conservation tillage fell from 86 percent to 47 percent even though
a maximum of 60 percent was permitted. In the Upper Mississippi high conservation tillage fell from 98 percent to 62 percent, less than the permitted maximum of 70 percent.

We consider run 2 of the ISU model to be more consistent than run 1
with our analysis of tillage technologies and our projections of total harvested land and its regional distribution.[19] Consequently, we consider
the amounts and regional distribution of erosion shown in run 2 of table
7-2 to be more consistent with our projections of production and harvested
land than the run 1 estimates.

Both runs indicate more sheet and rill erosion in 2010 than in 1977
(compare tables 7-1 and 7-2), both in total and per acre of cropland.
Unless wind erosion declines significantly, the projections imply an increase in total erosion. We see no reason to expect a decline in wind
erosion, given our projected increases in harvested land in the Northern
and Southern Plains and the Mountain States, the regions most affected by
wind erosion. Wind erosion is particularly severe in the Southern Plains
(see table 7-1), and our projections for that region show an increase in
land in main crops of 15.5 million acres (57 percent) between 1977 and
2010.

These results strongly suggest that if production and yields behave
as we have projected them, there will be considerably more erosion from
cropland in the United States in 2010 than in 1977, even with a significant
expansion of the amount of land in conservation tillage. We concluded

---

[19]As noted in the text, the model generated 17 million fewer acres of
harvested land in run 2 than in run 1. This was because in run 2 the model
was free to allocate production among regions while run 1 was constrained
by our original regional allocation. In the time between the two runs,
we revised our initial projections of production and yields, resulting in
a lower projection of land in main crops, hay, and fallow. Our revised
projection of harvested land was almost the same as the corresponding
projection in run 2: 417 million acres compared with 423 million acres.
We indicated in chapter 4 that we modified our initial allocations of
regional production and land use to bring them closer to the allocation
in run 2. Consequently our final projections of both total harvested land
and its regional distribution are closer to those of run 2 than of run 1.

earlier that current and future costs of present amounts of erosion are
cause for concern. If erosion mounts as our projections indicate, the
problem will become more severe, especially if the situation which emerges
is like that depicted in run 2.

We have no basis for judging the precise effects of the projected
levels of erosion on productivity of the land, but they surely imply that
productivity losses would be greater than those resulting from present
erosion. Nor are we able to estimate the off-farm damages of the projected
amounts of erosion. We do have estimates, however, of the amounts of
sediment delivered to water bodies around the country if erosion increases
as our projections indicate. These are shown in table 7-3, with corresponding estimates of sediment delivered in 1977. The estimates were made
by multiplying gross erosion in each year by estimates of sediment delivery ratios.[20] The estimates of sediment delivery ratios are very tenuous so the estimates in table 7-3 of sediment delivered should be taken as no
more than rough approximations. Even as such, however, they leave little
doubt that if erosion increases by 2010 as we have projected, the sediment
load in the nation's rivers, streams, lakes, reservoirs, and harbors would
increase sharply. Indeed, table 7-3 indicates it would almost double.
The Corn Belt would continue to contribute the greatest amount of sediment, followed by the Southern Plains. The greatest relative increases,
all in excess of 100 percent, would be in Appalachia, the Southeast, the
Mississippi Delta, the Southern Plains, and the Pacific Coast. (The absolute amount in the latter region would be minor, however.) This pattern
of regional environmental impacts is much the same as we found with respect to nitrogen fertilizers; and for essentially the same reason. These
are the regions with the largest relative projected increases in crop production and harvested land.

Should amounts of sediment delivered increase as shown in table 7-3
the quality of the nation's surface waters surely would be worse than at
present. How serious the deterioration would be perceived to be we are

---

[20]As indicated in the note to table 7-3, the source of the estimates
of sediment delivered and sediment delivery ratios is Gianessi, Peskin,
and Poles.

Table 7-3. Sediment Delivered to Water Bodies in the United States
(million tons)

Region	1977	2010 (Run 2)
Northeast	26.6	39.2
Lake States	45.7	93.0
Corn Belt	250.9	319.9
Northern Plains	141.7	246.5
Appalachia	57.7	139.4
Southeast	41.9	101.2
Delta	73.8	175.1
Southern Plains	69.3	256.4
Mountain	29.5	48.3
Pacific	11.2	30.8
Nation	748.3	1,449.8

Source: Data prepared for this study by Leonard P. Gianessi of Resources for the Future, based on data and procedures developed in Gianessi, Peskin and Poles (1980). Sediment delivered in each region is estimated by multiplying gross erosion in each region by estimates of sediment
delivery ratios.

unable to say, but we suspect it would be considered of significant national concern, justifying firmer measures to deal with erosion than any
previously adopted.

Conclusions

Our judgment is that the projections to 2010 of crop production and
land use do not imply an increasing problem of insecticide pollution. Indeed, it appears that changes in the location of cotton production and
innovation in the management of pests of cotton and corn will significantly
reduce the quantities of insecticides used. Moreover, the shift toward
less persistent insecticides should reduce damages to water quality. Although these materials are more acutely toxic to humans than the persistent insecticides, management of their environmental impacts should be

easier. We believe, therefore, that the insecticide problem will be of diminishing importance so far as the environment is concerned.

Given present and foreseeable weed management practices, our projections imply significant growth in the use of herbicides because of the projected increases in both cropland and conservation tillage. Present evidence does not demonstrate that the projected increases in herbicides applied will exact serious environmental damage. There are gaps in the evidence, however, relative to possible damages with present quantities of herbicides, and there may be scale effects that would emerge with the sharply larger quantities projected for the future. For these reasons there is no ground for complacency about prospective environmental damages of herbicides.

Soil and water salinity likely will continue to be a problem in the arid West, even though the amount of irrigated land in the region is likely to grow only modestly beyond present levels. However, presently known techniques for dealing with salinity, e.g., improved water and agronomic management, breeding more salt-tolerant varieties, appear sufficient to prevent the problem from becoming significantly worse.

If planned measures to reduce municipal discharges of phosphorus are in fact taken, then the projected increases in phosphorus fertilizers should not make the eutrophication problem more serious than it is at present. The projected increases in nitrogen in the Southeast, Delta, and Southern Plains, however, are relatively large and may contribute to the build-up of water quality problems in those regions.

The most serious threat to water quality by far, however, appears to be from projected amounts of erosion and sediment. While we cannot attach a value to the threat, a doubling of the amount of sediment delivered to the nation's surface waters surely would be cause for concern. When the effects of erosion in reducing the productivity of the land are taken into account, erosion, in our judgment, is easily the major threat to the nation's environment posed by the projected levels of crop production and land use.

References

Aldrich, Samuel. 1976. "Perspectives on Nitrogen in Agriculture: Food Production and Environmental Implications," paper presented at the annual meeting of the American Association for the Advancement of Science.

_____. 1980. Nitrogen in Relation to Food, Environment and Energy (Champaign-Urbana, Illinois Agricultural Experiment Station, University of Illinois).

Ayers, R.S. 1978. "A Case Study--Nitrates in the Upper Santa Ana River Basin in Relation to Groundwater Pollution," in P.F. Pratt (ed.), National Conference on Management of Nitrogen in Irrigated Agriculture (Riverside, University of California).

Baker, J.L., and H.P. Johnson. 1977. "Tillage System Effects on Runoff Water Quality" (no. 77-2504), paper presented at the winter meeting of the American Society of Agricultural Engineers.

Barisas, S.G., J.L. Baker, H.P. Johnson, and J.M. Laflen. 1978. "Effect of Tillage Systems on Runoff Losses of Nutrients," journal paper no. J-8534 of the Iowa Agriculture and Home Economics Experiment Station, Ames, Iowa, Project 1853.

Barnes, J.M., L. Heffner, R.F. Holt, R.G. Menzel, and R.D. Wauchope. (no date). "Influencing of Agricultural, Silvicultural, and Land Conversion Activities to Help Meet State and National Water Quality Objectives," SEA-USDA draft response to RCA. For information, contact R.D. Wauchope, USDA, Southern Weed Science Laboratory, Stoneville, Miss. 38776.

Cady, J.G. 1980. "The Natural Background: Soil Formation and Erosion," paper presented at Workshop on Soil Transformation and Productivity, sponsored by the Commission on Natural Resources, National Research Council, Oct. 16-17, 1980.

Castle, E.N., and I. Hoch. 1982. "Farm Real Estate Price Components, 1920-78," American Journal of Agricultural Economics vol. 64, no. 1 (Feb.), pp. 8-18.

Chemical and Engineering News. 1980. "Agent Orange Health Issues Raised Again," August 11.

Council on Environmental Quality. 1979. Environmental Quality, the Tenth Annual Report of the CEQ (Washington, D.C., GPO).

_____. 1981. Contamination of Ground Water by Toxic Organic Chemicals (Washington, D.C., GPO, January).

Crosson, Pierre. 1981. Conservation Tillage and Conventional Tillage: A Comparative Assessment (Ankeny, Iowa, The Soil Conservation Society of America).

Crosson, Pierre, and Kenneth Frederick. 1977. The World Food Situation: Resource and Environmental Issues for the Developing Countries and the United States (Washington, D.C., Resources for the Future).

Eichers, T.R., P.A. Andrilenas, and T.W. Anderson. 1978. Farmers' Use of Pesticides in 1976. USDA, Agricultural Economics Report No. 418.

Eijsackers, H. and J. van der Drift. 1976. "Effects on the Soil Fauna," in L.J. Andus (ed.), Herbicides: Physiology, Biochemistry, Ecology (New York, Academic Press).

Environmental Protection Agency. 1977a. National Water Quality Inventory, 1976 Report to Congress (Washington, D.C., EPA, Office of Water Planning and Standards).

_____. 1977b. The Report to Congress: Waste Disposal Practices and Their Effects on Ground Water, Executive Summary (Washington, D.C., EPA).

Frederick, K. 1982. Water for Western Agriculture (Washington, D.C., Resources for the Future).

Gianessi, L.P., H.M. Peskin, and J.S. Poles. 1980. "Cropland Soil Erosion and Sediment Discharge to Waterways in the United States," a report prepared by RFF for the U.S. Soil Conservation Service and the U.S. Environmental Protection Agency (Washington, D.C., EPA).

Giere, J.P., K.M. Johnson, and J.H. Perkins. 1980. "Is No-Till Farming the Answer to Soil Erosion?" Environment vol. 22, no. 6.

Goldberg, M. 1970. "Sources of Nitrogen in Water Supplies," in Agricultural Practices and Water Quality, proceedings of a Conference Concerning the Role of Agriculture in Clean Water held at Iowa State University, Ames.

Greaves, M.P. 1979. "Long-Term Effects of Herbicides on Soil Microorganisms," Annals of Applied Biology vol. 91, no. 1, pp. 129-131.

Hinish, W.W. 1980. "Soil Fertility," Crops and Soils Magazine (January).

Holt, R.F., H.P. Johnson, and L.L. McDowell. 1973. "Surface Water Quality," in Conservation Tillage, Proceedings of a National Conference (Ankeny, Iowa, Soil Conservation Society of America).

International Minerals and Chemical Corporation. (no date). "Fertilizer: the Indispensable Profit Maker," a report by the Marketing Services Department (Mundelein, Ill.).

Johnson, H.P., J.L. Baker, W.D. Shrader, and J.M. Laflen. 1977. "Tillage System Effects on Runoff Water Quality: Sediments and Nutrients," paper given at the meeting of the American Society of Agricultural Engineers, Chicago, Dec. 13-16.

Luhrs, C.E. 1973. "Human Health," in Nitrogenous Compounds in the Environment, EPA-SAB-730001 (Washington, D.C., EPA, Hazardous Materials Advisory Committee, 1973).

Lund, L.J., J.C. Ryden, R.J. Miller, A.E. Laag, and W.E. Bendixen. 1978. "Nitrogen Balances for the Santa Maria Valley," in P.F. Pratt (ed.), National Conference on Management of Nitrogen in Irrigated Agriculture (Riverside, University of California).

Martin, W.P., W.E. Fenster, and L.D. Hanson. 1970. "Fertilizer Management for Pollution Control," in Agricultural Practices and Water
Quality, proceedings of a Conference Concerning the Role of Agriculture in Clean Water (Ames, Iowa State University).

Meister, A.D., E.O. Heady, K.J. Nichol, and R.W. Strohbehn. 1976. U.S.
Agricultural Production in Relation to Alternative Water, Environmental and Export Policies, Card Report 65 (Ames, Iowa State University Center for Agricultural and Rural Development).

Miller, D.W., F.A. DeLuca, and T.L. Tessier. 1974. "Groundwater Contamination in the Northeast States" (Washington, D.C., EPA, Office of
Research and Development).

Mrak, E.M. 1969. Report to the Secretary's Commission on Pesticides and
Their Relationship to Environmental Health, U.S. Department of Health,
Education and Welfare (Washington, D.C., GPO).

National Academy of Sciences. 1975a. Pest Control: An Assessment of
Present and Alternative Technologies, vol. I (Washington, D.C., NAS).

_____. 1975b. Agricultural Production Efficiency (Washington, D.C.,
NAS).

Patrick, Ruth. 1973. "Aquatic Systems," in Nitrogenous Compounds in the
Environment, EPA-SAB-730001 (Washington, D.C., EPA, Hazardous Materials Advisory Committee).

Pimentel, D. 1971. Ecological Effects of Pesticides on Nontarget Species
(Washington, D.C., GPO).

Pimentel, D., E.C., Terhune, R. Dyson-Hudson, S. Rochereau, R. Samis, E.A.
Smith, D. Denman, D. Reifschneider, and M. Shepard. 1976. "Land
Degradation: Effects on Food and Energy Resources," Science vol.
194 (October 8) pp. 149-155.

Pimentel, D., D. Andow, R. Dyson-Hudson, D. Gallahan, S. Jacobson, M.
Irish, S. Kroop, A. Moss, I. Schreiner, M. Shepard, T. Thompson, B.
Vinzart. 1980. "Environmental and Social Costs of Pesticides: A
Preliminary Assessment," OIKOS vol. 34, no. 2, pp. 126-140.

Plewa, M.J, and J.M. Gentile. 1976. "Mutagenicity of Atrazine: A Maiz-
Microbe Bioassay," Mutation Research vol. 38, pp. 287-292.

Science News. 1973. "The Worse for Biodegradation," vol. 114, no. 13
(September).

Sommers, L.E., and V.O. Biederbeck. 1973. "Tillage Management Principles:
Soil Microorganisms," in Conservation Tillage, The Proceedings of a
National Conference (Ankeny, Iowa, Soil Conservation Society of Amer-
ica).

U.S. Department of Agriculture. 1974. Our Land and Water Resources
(Washington, D.C., USDA).

_____. 1979. "SCS Answers Critics of New Soil and Water Data,"
Office of Governmental and Public Affairs, Issue Briefing Paper no.
20 (Washington, D.C., USDA, August 21).

U.S. Department of Agriculture. 1980a. Basic Statistics, 1977 National Resources Inventory (Washington, D.C., Soil Conservation Service, February).

_____. 1980b. RCA Appraisal 1980 Review Draft, Parts I and II (Washington, D.C., USDA).

U.S. Department of Interior. 1975. Westwide Study Report on Critical Water Problems Facing the Eleven Western States (Washington, D.C., Dept. of Interior, April).

U.S. Department of Interior, and U.S. Department of Agriculture. 1977. Final Environmental Statement: Colorado River Water Quality Improvement Program, vol. I, section 1 (Washington, D.C., Bureau of Reclamation).

U.S. Department of Interior, California Department of Water Resources, and California State Water Resources Control Board. 1979. Agricultural Drainage and Salt Management in the San Joaquin Valley, Final Report (Fresno, Calif., San Joaquin Valley Interagency Drainage Program).

U.S. Water Resources Council. 1978. The Nation's Water Resources, 1975-2000; Second National Water Assessment (Washington, D.C., WRC, December).

van Schilfgaarde, Jan. 1977. "Minimizing Salt in Return Flow by Improving Irrigation Efficiency," in J. Law and G.V. Skogerboe, Proceedings: National Conference on Irrigation Return Flow Quality Management (Fort Collins, Colorado State University).

von Rumker, Rosemarie, E.W. Lawless, and A.F. Meiners, with K.A. Lawrence, G.L. Kelso, and F. Horay. 1975. Production, Distribution, Use, and Environmental Impact Potential of Selected Pesticides, prepared for the Environmental Protection Agency's Office of Pesticide Programs, and Council on Environmental Quality (Washington, D.C., EPA).

Wauchope, R.D. 1978. "The Pesticide Content of Surface Water Draining from Agricultural Fields--A Review," Journal of Environmental Quality vol. 7, no. 4 (October-December).

Wauchope, R.D., L.L. McDowell, and J. Hagen. no date. "Environmental Effects of Limited Tillage," in Weed Control in Limited Tillage Systems (Champaign, Ill., Weed Science Society of America), in press.

Wischmeier, W.W., and D.D. Smith. 1965 (reprinted 1972). Predicting Rainfall-Erosion Losses from Cropland East of the Rocky Mountains, Agricultural Handbook no. 282 (Washington, D.C., USDA-Agricultural Research Service).

Chapter 8

POLICY CONSIDERATIONS

Background

The previous chapter showed that expanding agricultural production promises to generate significant new environmental stress, with implications that extend beyond the farm. Policies to cope with that stress will be set in the context of the long record of agricultural policy, but they also must recognize changing circumstances and values outside of agriculture that impinge upon it. A brief review of this changing context is needed.

Agricultural policy and agricultural land policy have been central to the political and social history of our republic. The availability of free or low-cost land was a primary impetus to early settlement and to the extension of the frontier. With land in ample supply, restrictions on its use were not considered. Fee simple ownership and the virtually unrestricted right to use owned property in any way desired were embedded early in the nation's institution and psyche. Thomas Jefferson's views on a democracy of landowners gave intellectual sanction to this system. Early debates over such matters as tariffs, transportation, banking and credit, policies for the disposition of federal land, and even slavery, all reflected the farmer's eagerness to acquire and work his own land and to sell his products on a wider market. A largely agrarian society of free men in which widespread access to land ownership played a central role was for decades our major dynamic and was a beacon to the world.

The need to adjust to changing competitive, resource, and environmental factors has been present from the beginning. Erosion and soil exhaustion in some of the early settled areas of the South could not be countered by techniques known at the time. Land was abandoned as opera-

tions, especially plantations, moved further west. In Appalachia, similar problems often made farming unviable and resulted in populations
stranded on unproductive land where they remained for generations. New
land to the west had better soils and climate that also made it difficult
for New England to compete, especially after improved transportation made
the West more accessible. Thus, regional shifts in agricultural production were an early consequence of both economic and technical factors that
were reinforced by policies affecting land disposition and transportation.

By the end of World War I (WWI), the center of gravity of American
agriculture had moved definitively to the Midwest, and agricultural policy
issues had begun to assume their long-term form. Ultimately, two broad
groups of policy issues predominated: (1) how could farmers be assisted to
hold onto their land and realize a decent income in the face of sharp
price fluctuations or chronic surpluses of production and the resulting
depressed prices? And (2) what assistance could government provide in
helping farmers maintain production on lands vulnerable to erosion and
drought? Often the two objectives merged with fragile lands being with-
drawn from crops under government support programs to protect both the
land and the market. Other land was provided with subsidized water under
conservation projects.

Prior to WWI, policy (except for land disposal policy and irrigation
projects) tended to operate at a general level through monetary, credit,
trade, and transportation measures. Thereafter a shift to the farm level
occurred, with policy now operating through government loans, support payments, farm conservation plans, and extension services. The once independent farmer has become directly dependent on this policy net and very
insistent and influential in the halls of Congress on behalf of public
support for the farming sector.

Distress in the countryside, especially among grain farmers, was
endemic following WWI. The farming sector entered a period of stress in
the 1920s and virtually collapsed in the following decade. What had been
an economic problem was now a major sociological issue as land and farmers
were compelled to leave farming while faced with poor opportunities elsewhere in the economy. After WWII, the move away from the farm accelerated, propelled now by a dramatic revolution in agricultural technology

that increased yields while greatly diminishing labor requirements. Unlike in pre-war years, the swelling cities and growing service economy
now could absorb the people displaced, but land and farm commodities still
remained in surplus.

During the decade of the 1970s, the major human adjustment of the
farm economy was substantially completed. The number of low income farmers had been reduced and the size of farm increased to the point where
rural poverty no longer was a dominant sociological problem. A series of
events--weather, a slower rate of yield increase, and strong foreign de-
mand--evaporated chronic surpluses, raised crop prices, and absorbed idled
cropland back into production. Familiarity with soil conservation techniques was now widespread in the countryside, and conservation practices
commonly were adopted when profitable to the farmer or if sufficiently
subsidized. Despite these changes, the structure of farm programs re-
tained much of the form acquired during the time of surplus; those programs, especially the price support structure, are jealously guarded by
farm interests and have become reflected in the value of farm enterprises.
All are oriented toward assisting the farmer. When they make him the
instrument of conservation policy, they are dependent on his voluntary
cooperation. However, with demand for farm products strong in the 1970s,
the once most effective income support and conservation practice--the
return of cropland to cover--lost much of its appeal.

Yet another important strand of agricultural policy has been the
farmer's continuing interest in the widest possible market for his products. Cotton exports once sustained the South, and grain shipments
built the mid-continent rail system. Exports were quiescent during the
protectionist years of the Great Depression, and farm policy focused on
domestic policies for averting surplus. With the return of foreign demand after WWII, driven especially in recent years by prosperity in developed countries and by both population and income growth elsewhere,
farmers have regained their keen interest in unrestricted exports. Government restrictions on the grain trade are strongly resisted.

The surviving structure of agricultural policy does not seem one
that would be suggested de novo for the conditions of today. Farmers
have now become substantial businessmen atuned to international markets.

They are no more deserving of price supports than are other sectors of
the economy that must compete in world markets. Meanwhile, a new environmental consciousness has been born that demands control of environmental
insult. It holds economic actors responsible for the off-site consequences
of their operations. Farm-related sediment, pesticide residues, and the
like no longer are viewed with indifference. At the same time, a new
spirit of conservation has arisen that shows concern about limits to the
nation's resources; it views preservation of the renewable productive
potential of the land as a matter of social interest that need not be left
solely in private hands, despite our long tradition of fee simple ownership. Thus, new concerns have been introduced into the policy arena that
are less farmer oriented. It is difficult to know which trends are ephemeral and which will prove enduring. Environmental programs face a somewhat more skeptical reception now than over the past decade. But concern
for soil conservation does not appear to have abated.

Other societies sometimes have neglected soil conservation and environmental matters to their own peril; failure to deal with these issues
on a timely basis has been a factor in the demise of some. However, in
its long denial of social responsibility for these matters that are of
concern to all, our own society has been somewhat of an aberration. In no
small part this goes back to our history of unfettered private ownership
of land and to the circumstances which made the system of private ownership both economically productive and politically liberating. Carried to
doctrinaire extremes, however, this viewpoint paralyzes the capacity of
society to deal with newly recognized and more acute problems of environmental control and conservation.

There is a national interest in the maintenance of an efficient and
productive agriculture that permits moderate food prices and generates
high export earnings. Sometimes it is difficult to express that interest
in ways that are environmentally benign, for that may conflict with the
private aims of farmers. While regulation can restrict the way in which
land is used by forbidding offending practices, its positive thrust is
weaker; it cannot compel private owners to employ land in socially beneficial ways when it is not profitable for them to do so. Therefore, a
policy to control the environmental consequences of expanded agricultural

production must acknowledge both the legal standing and managerial role of farmers and must not threaten the maintenance of a viable private agriculture. Realistically, policy also must acknowledge the ingrained policy expectations of farmers, for these expectations make abrupt or drastic change politically difficult.

Current Environmental Approaches

Existing programs for soil conservation, water pollution control, and pesticide regulation differ greatly in origin and form and are administered by different agencies. The principal purpose of soil conservation programs is to preserve and enhance the productive potential of the soil. At times, such programs also have become a vehicle for farm income maintenance or for commodity price stabilization, though that is not currently the case. The programs, administered by U.S. Department of Agriculture, are overwhelmingly voluntary.

Water pollution control programs are administered by the Environmental Protection Agency but, with respect to agriculturally-related nonpoint source pollution, the agency delegates nearly all planning and control to the states through its program under Section 208 of the Federal Water Pollution Control Act (Clean Water Act). In practice, 208 program have tended to become extensions of present soil conservation programs.

Pesticides are regulated in three ways: through the registration process administered by EPA, through Food and Drug Administration, and through USDA rules on pesticide application procedures.

The other major concern of a resource and environmental nature is the preservation of agricultural land. While USDA and EPA have policy positions in favor of land preservation, most initiative in this field lies with the states where it becomes entangled with other nonagricultural land use considerations. Finally, various programs for wetland preservation, upstream flood control, and fish and wildlife habitat have implications for agricultural land use that may be restrictive in character.

Are existing approaches well adapted to meet current needs and the prospective needs suggested by our projections? In order to answer that question each policy area must be examined in order, but it is apparent

that the disparate programs do not adequately consider their interrela-
tionships. For example, conservation tillage is a most promising means
for reducing erosion, but it implies greater use of pesticides and will
be feasible only if they remain available. Their use in turn may aggra-
vate water pollution problems and damage valued nontarget species.

Soil Conservation Programs

Three principal federal soil conservation programs operate at present. The Soil Conservation Service (SCS) Conservation Operations Program
provides technical advice and conservation plans to farmers. The Agricultural Conservation Program is a means for channeling money to farmers to
encourage them to carry out conservation practices on their land. It is
administered by the Agricultural Stabilization and Conservation Service
(ASCS) in cooperation with local committees of farmers. The Great Plains
Conservation Program is a special effort by SCS to encourage farmers in
that area to contract for an agreed operational plan that will reduce wind
and other erosion from cropland and range.

The SCS Conservation Operations Program is strictly voluntary. Its
purpose is to provide technical assistance to farmers. The initiative
lies with the farmer, who must request assistance from SCS. The SCS representative prepares a conservation plan for the farm that incorporates a
mix of farming practices or the installation of more permanent features,
such as terraces.

A review of this program by the General Accounting Office (GAO) in
1977 criticized it as passive in not seeking out and concentrating on land
with the most severe erosion problems (GAO, 1977). Moreover, the GAO
report argued that many of the plans were too elaborate, were not used by
farmers, there was little follow-up by SCS, and that the soil loss results
for those cooperating with SCS were little different from those who did
not. SCS acknowledged the validity of these criticisms. But SCS plays a
wider educational role in cooperating with Soil Conservation District
members and in other ways, so its impact probably is somewhat greater
than the GAO report admitted. The program has enlisted over one-half of
all farm units as cooperators. If conservation has not attained its
goals, that can hardly be attributed to widespread unawareness of good
conservation practice in the countryside.

Soil Conservation Districts (SCDs) or related units play an important role. These units were created under state legislation to promote conservation. Their role is in all cases educational, but in some states they have added powers of taxation and enforcement that permit them to do some planning and to build control structures of wider benefit.

The Agricultural Conservation Program is a cost-sharing program to make funds available to farmers for conservation practices. The program is under the jurisdiction of ASCS and operates through state and local committees that make the actual allocations of funds. SCS provides technical advice to the committees and to farmers carrying out the practices. Again, the program is entirely voluntary. The major objective of the program is to secure enduring soil and water conservation practices that the farmer would not find profitable if financed from his own funds. The ASCS argues that it has broader responsibility for the protection of woodlands and wildlife as well as water conservation and pollution control, so it should not be held narrowly to the funding of erosion control measures.

In reviewing this program, the GAO found that it was not as strongly focused on soil conservation as would be desirable; over one-half of the available funds went for other measures such as tile drainage, irrigation, land leveling, and liming that were primarily in the farmer's own interests. Moreover, there appears to be a tendency on the part of local committees to foster those conservation practices that are enduring--i.e. structural works--over such nonstructural practices as conservation tillage or rotation systems. The latter could be promoted by term contracts, but this device has not been popular. As a consequence of programs designed by Congress, the ASCS has a split personality. In one program it promotes conservation, which might require the retirement of vulnerable acreage, while in another it offers price supports based on historical acreage--a system that penalizes premature land retirement.

A subsequent ASCS evaluation of ACP supports much of the GAO criticism of this program (USDA, 1980). Their examination of recent experience on agricultural lands (not just cropland) found that the acreage on which uncontrolled erosion exceeds 5 tons per acre per year received only 48 percent of all assistance in erosion control practices. Conservation tillage, one of the most cost-effective practices for erosion control, was the practice least often assisted (USDA-ASCS, 1980).

The Great Plains Conservation Program, administered by SCS, is an effort to induce farmers to change cropping systems and land use to conserve soil and water in that region. The chief practice to accomplish this is the provision of permanent cover on lands especially vulnerable to erosion. The chosen policy instrument is long-term contracts in which the farmer agrees to follow conservation practices in return for government payments. However, when reviewed by the GAO, the program had been implemented on only one-fourth of the intended acreage and had little prospect of reaching beyond half of it by 1981, the terminal year of the program. Strong grain prices induced farmers to return land to cultivation when contracts expired, or to decline those offered. As with other programs, there was little evidence that priority was being given to the most affected acres or to the most effective practices.

Water Pollution Control

Agriculturally related water pollution is subject to various provisions of the Clean Water Act administered by EPA. Identifiable point sources, such as large cattle feedlots, are controlled through a permit system that specifies allowable operating practices. Most agricultural pollution does not come from point sources but rather in runoff from fields. Control of nonpoint pollution is delegated to the states, subject to the approval of state control plans by EPA. Under Section 208(b)(2)(F) of the Clean Water Act, states are required to plan for areawide waste treatment management, and the plan is to include "a process to (i) identify, if appropriate, agriculturally and silviculturally related nonpoint sources of pollution, including return flows from irrigated agriculture, and their accumulative effects, runoff from manure disposal areas, and from land used for livestock and crop production, and (ii) set forth procedures and methods (including land use requirements) to control to the extent feasible such sources." The state 208 planning effort thus set in motion has been supported by federal grants administered by EPA. Completion of the plans was laggard, but the first round is essentially done and most plans have been approved by EPA.

While EPA provided financial assistance and administrative guidelines for the preparation of 208 plans, it allowed states great latitude in their preparation. In particular, EPA has shunned the assertion of any regulatory role for the agency, even though it has authority for it, and it has encouraged states to favor voluntary approaches as well. The emphasis is on measures known as Best Management Practices (BMPs) that can be adopted by individual farmers on their land to prevent pollution. In most instances, the BMPs prove to be familiar soil and water conservation measures to control erosion and runoff, though there also is some emphasis on those measures that inhibit sediment delivery to streams once erosion has occurred. The plans, being relatively new and mostly voluntary, have not compiled any real record of implementation to date. However, most contemplate making use of existing administrative delivery systems for conservation programs in the countryside; i.e., they will work through the Soil Conservation Districts and Resource Conservation Districts to contact farmers, and SCS personnel will provide technical help. States may invest their own financial and administrative resources in these programs as well, though few do so on any important scale.

In an effort to give impetus to agriculturally-related nonpoint source control, the Rural Clean Water Program (RCWP) was established by amendment to Section 208 to provide for a program of long-term (three to ten year) cost sharing contracts for the installation of BMPs to improve water quality. The aim of this provision is to target more directly on water pollution control than is the case with other conservation programs. Responsibility for the program is shared between EPA and the USDA, with the latter having control over field implementation but sharing project approval and the approval of BMPs with EPA.

Funds are available to farmers who are operating under a water quality plan approved by their SCD. Local committees determine priority among farmers for funding. Areas and pollution sources being targeted are those with the most significant effect on water quality, so the geographical scope of this program is restricted to critical areas and sources. As of 1980, thirteen projects were being funded across the country. For a project area to qualify under the program farmer participation must be on the order of 75 percent (USDA, March 1980).

The concept of this program is very promising. It is rather sharply focused on areas with severe quality problems. The projects must be in conformance with 208 plans if they are to gain approval, and many states have looked to RCWP as an important tool for eventual implantation of their plans. Since RCWP provides for cultural as well as structural practices, cost effective measures such as conservation tillage and the maintenance of border strips can be encouraged. The program shuns measures intended to enhance production in favor of those affecting water quality that farmers would not undertake on their own because they are not profitable. The scale of the program is very modest, with initial funding established at $50 million per year. The program is referred to as experimental, and no operating results were available at the time of writing. However, it is clear that funding at this level will not make a major impact on farm-related water pollution nationally.

EPA has provided 208 planning grants and guidelines, reviewed the plans submitted by states, has made some effort to assess the character and severity of the problem nationally, and has approved lists of BMPs to be followed in controlling water pollution. Except for this, most of the responsibility and initiative has fallen to the states to control farm-related nonpoint pollution. Their performance has been extremely variable, but most often not deserving of praise. Some states had to start from scratch to establish planning staffs. In many cases they viewed the task as one imposed upon them by the federal government, and it was not necessarily a response to a problem that they felt deserved attention. Initially, there was concern that the federal government would use the law to gain sweeping control over land use in the countryside. That fear has receded, but meanwhile soil conservation, including its water quality aspects, has assumed a more prominent place on the national agenda, and latent concern about a major federal role may yet be justified.

The authors examined 208 planning in six representative states to learn how it had been conducted: Arkansas, California, Georgia, Iowa, Nebraska, and Texas. They were chosen as representative of major farming areas--not because they are examples of either good or bad water quality or planning. However, Iowa led the nation in sheet and rill erosion and sediment delivered in 1977, and Texas led in both wind erosion and total erosion. Data showing gross sheet and rill erosion and sediment delivered in each state in 1977 and as projected in 2010 are shown in table 8-1.

Table 8-1. Estimates of Gross Sheet and Rill Erosion and Sediment Delivery from Cropland for the United States and Selected States, 1977–2010

	1977		2010 Run 1		2010 Run 2	
	Gross Erosion[a] (000 tons)	Sediment[b] (000 tons)	Gross Erosion[b] (000 tons)	Sediment[b] (000 tons)	Gross Erosion[b] (000 tons)	Sediment[b] (000 tons)
United States	1,925,849	765,113	2,320,210	1,001,110	3,537,031	1,486,418
Arkansas	46,711	22,421	36,572	17,555	68,686	32,969
California	8,591	3,386	6,125	2,389	6,190	2,414
Georgia	42,662	14,777	83,445	29,205	92,729	32,455
Iowa	261,253	92,338	104,752	36,663	253,276	88,647
Nebraska	117,792	46,871	49,100	19,640	130,854	52,341
Texas	99,546	52,539	324,924	172,210	392,971	208,275

Sources: [a] U.S. Department of Agriculture, Basic Statistics, 1977 National Resources Inventory (Washington, D.C., February 1980) table 16.

[b] Provided by Leonard P. Gianessi, Resources for the Future. Runs 1 and 2 correspond to those discussed in chapter 7. See especially tables 7-2 and 7-3.

In conducting our survey of 208 planning in the six states, we interviewed responsible officials and knowledgeable academics in each case, and collected documents wherever available. The survey was done in 1979-80.

Iowa is the preeminent Corn Belt state, with an intensive, productive, and sophisticated agriculture. Corn and soybeans are erosive crops and that, together with the rolling terrain in much of Iowa, and other factors, has caused that state to lead all others in water erosion from cropland. Concern about this problem brought state legislation in 1971 establishing soil conservancy districts and defining maximum allowable soil loss limits on agricultural land comparable to SCS tolerance levels. The law has not been very effective. The state cost sharing envisaged has been minimal. No mandatory action can be taken against violators unless state cost sharing funds are available to help correct the problem. So far nearly all of the small state appropriation has gone to voluntary participants (EPA, 1979). Moreover doubts have been raised about the wisdom of overly vigorous enforcement of the law because it would place Iowa at a competitive disadvantage vis-à-vis other states (Nagadevara, Heady, and Nicol, 1975).

The widespread concern about soil conservation has tended to overshadow water quality measures and has presented some real dilemmas. As noted in the last chapter, conservation tillage reduces erosion but may increase run-off of chemicals. The widespread use of tile drains that carry much nitrogen leached through the top layers of the soil further contaminates water in Iowa. Nonetheless, most attention has gone to erosion and sediment reduction by means of the usual conservation measures (Iowa State University College of Agriculture, 1978).

While the governor affirmed that Iowa had a nonpoint water quality problem, very little assessment of its severity was done. The approach was more subjective. The aquatic life potential of the water was taken as a standard and then an attempt was made to identify the pollution sources that prevent attainment of that standard or of the standard of body contact for recreational use.

With regard to 208 implementation, the plan is mostly voluntary in its approach. The intention is to work with those who show a willingness to participate rather than to apply pressure on the worst polluters. Any mandatory action is to be left to the conservancy districts. Great hope

was placed on RCWP funds at the outset--a hope that now seems largely
misplaced.

While Iowans seem acutely aware of water quality problems stemming
from agriculture and are disposed to act on them, elsewhere there often
has been much less attention to the problem. In neighboring Nebraska,
where an expanding irrigated agriculture faces rising energy costs, there
was much less attention to water quality and erosion. While a 208 plan
was under preparation, the effort encountered great skepticism and fre-
quent questions about the need for it.

Perhaps because of that, unusual emphasis was placed on public participation. This even extended to problem identification. Local advisory
committee members were invited to vote on which water quality problems
they considered deserving of highest priority. They rated soil erosion
as the top priority problem, center pivot irrigation on marginal lands
as second, and leaching and runoff of fertilizer and chemicals as fourth
and fifth. This hardly impresses as a very scientific procedure and the
planning agency acknowledged the need for further refinement of the list.
Discussions with academic experts supported the importance of erosion and
leaching, but there was agreement that sediment is not the problem in
Nebraska that it is in Iowa. Many people, including some of those responsible for 208 planning, saw little need for planning, in part because many
small rural streams have few beneficial uses in any case. However, contamination of groundwater was acknowledged to be a problem in a state that
relies very heavily on underground supplies.

In view of widespread doubts about the need for 208 planning in Nebraska, planners firmly intend to stress voluntary measures for implementation. Those which would be in the farmer's best interests will be promoted. Thus, control of irrigation runoff where nutrients are applied
with the water can save the farmer money and should appeal to him. Contour farming and chisel plowing already are accepted in many parts of the
state.

Nebraska is the site of one of a half-dozen USDA-EPA model implementation projects (MIPS) where a concentrated attack on rural water quality
problems is being attempted on a limited acreage. The problem confronted
in this instance is severe erosion from straight row corn cultivation in

hilly terrain. Expensive and heavily subsidized terraces are being built on some farms to the neglect of more cost effective contouring and conservation tillage. Farmers in the area are not strongly motivated to arrest erosion because they have deep soils. Even the water quality benefits seem uncertain, for the small local streams have limited use at best. It is not clear what the project will demonstrate. While such concentrated effort may succeed technically, it is too costly for widespread application.

Texas attempted a very decentralized approach to 208 planning, as befits a state of its size and diversity. The state's Department of Water Resources has primary responsibility for problem identification, which it undertakes by river basin. Although streams in Texas generally carry a heavy sediment load, the state has no standard for sediment and does not treat it as a pollutant. Sediment is attributed to natural conditions and is not thought to be closely related to agriculture. In the High Plains area there are no streams, and drainage is to evaporative playas (dried lake basins). Sheet and rill erosion is not severe on the rolling plains and is moderate on the black plains; there seems little evidence that it affects productivity significantly in those areas.

Erosion in Texas is mostly by wind, and the amount (14.9 tons per acre of cropland per year—table 7-1) is well in excess of "T" values set by the SCS. Wind erosion evidently is not viewed as a serious threat to water quality; in any event it does not fall within the purview of 208 planning in Texas, or anywhere else. To the extent that wind erosion in Texas is addressed, it is primarily through the SCS's Great Plains Conservation Program. The limitations of that program, as assessed by the GAO, were pointed out above in the section on soil conservation programs.

Water pollution by agricultural chemicals is not regarded as a major problem by 208 planners in Texas. Some leaching of chemicals to groundwater is noted in the winter garden section of the state. In the coastal rice growing areas, a heavy rain just after the application of pesticides may wash enough into bays and estuaries to affect shellfish, though this point is disputed.

The 208 screening process in Texas examined over twenty areas where agriculture was suspected of contributing to failure to meet stream standards, but only three or four involving arsenic, nitrate concentrations, and fecal coliforms were finally attributed to agriculture. The State Soil and Water Conservation Board, an overgroup of conservation districts, has responsibility for developing plans to cope with the identified problem areas. They in turn work with local conservation districts in that process. Land use surveys and the designation of appropriate BMPs are the basis for plans in identified problem areas, and the plans were seen as vehicles for seeking RCWP money for Texas. The conclusion must be that most Texas agricultural land is not affected by 208 planning and the state has minimal enthusiasm for the process. Normal soil and water conservation programs continue, with stress on wind erosion in the High Plains and on contour plowing and water retention in other parts of the state.

Arkansas planners put their main emphasis on developing an information system that should allow them to identify sources of nonpoint pollution. They sought to develop detail on gross erosion and to apply specially tailored sediment delivery ratios for each small watershed. Nutrient and pesticide loads were derived from the recommended application rates for these materials. Identified problem areas are to be more intensively studied. Arkansas has much high quality water in the northwest part of the state where agricultural pressure on water quality is lower than in the more intensively farmed southeastern delta area.

Problem identification appeared to exhaust the energy of the planners at the time their state was examined. No statewide plan for implementation was contemplated, and they foresaw little other than an educational approach with voluntary compliance as likely. It is conceded that streams in rice growing areas do not meet standards and are not likely to; little use is made of them for fish or recreation, so there is a tendency to accommodate the convenience of farmers.

Georgia's approach to 208 planning has been extremely cautious. The responsible state planning agency notes that there has been little stream monitoring in agricultural areas, so they cannot document the link between water quality and agricultural activities. Instead they have identified potential problems by assembling data on land use characteristics by county and watershed. From this an erosion potential was computed and a sediment delivery ratio applied so as to rank counties by sediment delivery potential. Sales of pesticides by county were used as an indicator of that problem. The inadequacies of this approach are obvious to planners, and they propose to study in more detail the links between potential problems as identified by land use and actual problems seen in water quality. They see sediment as their primary pollutant.

Implementation of the Georgia plan will be left to an overgroup of soil conservation districts who propose to use education and demonstration projects to encourage compliance. While the state has authority to regulate, it does not feel that it has established a basis for doing so or that there would be political support for it. Indeed, there is little support for a regulatory approach anywhere in the Southeast, although other states have been more specific than Georgia in identifying problems. Farmers are not convinced that the problems are clearly enough identified and serious enough to warrant it, and state officials appear to agree. Established state stream standards may be met in Georgia, even though the larger goal of "fishable and swimmable" water is not.

California has a long history of water planning, including attention to water quality. The state's agriculture and physical conditions are too diverse to allow a single approach to quality management. Much agriculture is conducted in metropolitan areas where its environmental effects are submerged with those of other nonpoint sources. Intensive irrigated agriculture predominates. While erosion is severe on some hillside orchards and grain fields, much of the land in crops is flat, and erosion is not the major problem in California. High concentrations of nitrate-nitrogen in groundwater in some areas were noted in the last chapter, as were the salinity problems in some fertile areas of the San Joaquin Valley.

Planning under Section 208 in California has consisted of updating already existing basin plans. This is done largely by the state's regional water resources control boards. No single document summarizes their work. The boards have responded to local concerns. Implementation again is left to local conservation districts in the first instance, but the state seems much more willing and able to intervene than in the other states examined. Under the state's Porter Cologne Act, they have considerable regulatory authority, many counties have local statutes that

can be invoked, and state funds are available to finance water control
programs (John Muir Institute, 1979). Moreover, agriculture in California
is highly organized, and farm groups are willing to bring pressure on
identified offenders in hopes of averting more general regulations affecting everyone. As a result, there was a greater sense of ability and willingness to cope with farm-related water pollution. However, state officials and others attribute little of this to 208 planning; rather it is
considered to be the product of state efforts that antedate and go beyond
208.

A review of 208 planning frequently disappoints the expectations
generated by federal law. Nonetheless, one should not discount what has
been accomplished. States have been forced to face farm generated water
problems explicitly and to build planning and administrative capacity to
deal with them. This can be an important step if the problems become more
aggravated, as our projections suggest they will. The states have had to
establish some kind of identification and priority system, however inadequate, and to give evidence of legal authority to act where they deem it
necessary. The result has been educational for the states and has led a
few to take initiatives to correct their problems without waiting on federal money. The subtle hint of enforcement that underlies the law has
made states, farm organizations, and farmers more self-conscious about
the need to act on their own. And the bait of RCWP funds has induced
serious planning for some key problem areas that may create its own momen-
tum for action, even though federal money is scarce.

Pesticide Controls

The use of pesticides in agriculture is regulated to prevent the contamination of human foods, to safeguard the health of agricultural workers,
and to protect nontarget species (including man) from excessive damage.
EPA has primary responsibility for pesticide controls, but other agencies,
especially the Food and Drug Administration and USDA, have assigned roles,
and states may also regulate to stricter standards if they choose.

Residues of pesticides in food may be directly toxic, or they may be
suspected as carcinogenic, mutagenic, or teratogenic. Acutely toxic doses
would not be expected on food. But because some pesticides are persistent

in the environment and are concentrated by biological processes and may accumulate in certain tissues and organs, the possibility of toxicity is present at very low dose rates. Moreover, chronic low level exposure to those chemicals that may be carcinogenic raises all of the murky issues about thresholds and the validity of animal experiments or epidemiological data that so bedevil all discussions of the etiology of cancer.

Responsibility for setting tolerance levels in food lies with EPA. They establish guidelines for the amount of each pesticide that is allowed in different foods. Where processing of raw products reduces the concentration of pesticide, attention goes to the processed product. If there is evidence that pesticides have not been used in accordance with the label, action may be taken against treated food products even if no tolerance levels have been exceeded. Actual enforcement of these regulations is left to the FDA, which monitors food products about the country, and to the USDA, which has responsibility for meat and poultry. Since agricultural commodities that do not meet standards are essentially unmarketable, farmers have a most powerful incentive to avoid contamination.

The heart of federal regulation of pesticides lies in the registration and classification process under the Federal Insecticide, Fungicide, and Rodenticide Act (FIFRA) as amended. Registration provides information about the contents and effectiveness of the material and specifies the uses for which it can be sold. In addition, EPA may classify pesticides according to whether they are for general or restricted use. In the latter case, they can be restricted to application by certified applicators only. In combination, these provisions are the major protection to man and to other nontarget species from pesticides in the environment. Human occupational exposure is further protected by rules of the Occupational Safety and Health Administration (OSHA). States also play a role in regulating occupational exposure or, in some circumstances, in registration.

The law has proved difficult to administer. The number of compounds on the market is so large (about 35,000) that it has defied attempts at careful evaluation of those already registered (NRC, 1980). As a consequence, EPA's power to suspend or cancel existing registrations has become the focal point of efforts to restrict use. Controversies over individual compounds and classes of pesticide abound, and the scientific basis for decision often is ambiguous. Nonetheless, important agricultural pesticides, starting with DDT, have been cancelled, suspended, or restricted in use.

Chemical companies complain that the cost of developing, testing, and registering or defending a new compound frequently is prohibitive. Since pests tend to develop resistance to insecticides, it is important that successive generations of pesticides become available. Any system of pest control, including IPM methods, will continue to make use of chemicals, even if in smaller amounts. This problem presents a real dilemma for policy. EPA has considerable latitude to interpret the law, and increasingly they have used risk/benefit analysis as a judgment tool (NRC, 1980). It is recognized that successful pest management cannot be conducted on a no-risk basis.

Progress has been made in the way in which chemicals are used, not only because of the law, but also because of greater environmental awareness. While the standards for the certification of applicators have not been stringent, a more responsible attitude toward pesticide application has spread to farmers and other users. And ever-vigilant citizen environmental groups have channels for complaint against careless users.

Modern agricultural technology has evolved into a dependence on pesticides that is not easily controlled by regulation. Some otherwise benign practices such as conservation tillage increase the need for pesticides. Technologies that would reduce that need can be designed, but generally they require a considerable lead time for development and their development may not offer the same commercial possibilities that broad spectrum chemicals do. Regulation to restrict present chemical use might stimulate alternatives if they were profitable for private firms. Since most alternatives—resistant varieties, attractants, pathogens, cultural practices, etc.—do not offer large-scale commercial opportunities, their development most often depends on publicly supported research. The regulatory program of EPA is not well coordinated with the provision of such alternatives and is weakened by their absence. Clearly government strategy in this field needs a longer view.

We conclude that our projections of output and resource use threaten to generate significantly greater environmental stress than current poli-

cies have been asked to cope with. The added stress will occur in geographical areas where current programs are weakest and in those areas
will show up mostly in the form of erosion losses and attendant water
pollution problems. It is important to consider alternatives to the
erosion control policies now in place.

Alternative Policy Approaches

The environmental consequences of expanding agricultural production
need not be met by environmental policies alone. A whole range of governmental policies do affect or could affect the scale, location, and conduct
of agriculture, with implications for its environmental results. Our
analysis suggests that the preeminent environmental threat from agricultural activities is erosion damage, certainly that off the farm and perhaps that on the farm in lost productivity. In principle, agriculture
should be conducted in a manner that holds soil loss to acceptable levels
so as to cover the full cost of production and if it does not then we may
wish to consider policies that restrict output. Restraints affecting demand would be one course, although our projections do not indicate any
near term need for this. There also appear to be ways to increase output
while controlling soil loss and abuse, but these may involve more intensive use of the best lands and therefore imply other environmental pressures. Such measures would likely change the character and location of
production and would affect both individual farmers and producing areas
quite differently.

Very likely demand will be left to find its own level and attention
will go to better coordination of production, conservation, and environmental policies to support one another. Water quality may be made a more
explicit concern of policy, and conservation and environmental policies
will be more vigorously pursued with programs focused more sharply on
problem areas. Other agricultural programs may be asked to lend support
to these objectives. The way in which farming is conducted then may receive more attention than its scale or location, but the result could
still be some reduction in output from what it otherwise would be.

Demand and Soil Degradation

In contemporary America, agricultural output expands in response to demand--to sales of farm commodities at attractive prices--and expanding output increases environmental pressures of all kinds, but especially the threat of erosion. This demand-driven output differs from that of decades past when individual farmers felt compelled by falling income to produce more but, recognizing the futility of that, collectively were willing to restrict output in return for income support. Conservation measures that idled land or converted it to less intensive use in that context were consistent with the policy of dampening output and removing resources from agriculture in order to improve the economic position of those operators who remained.

Now the pressures are all in the other direction: farmers see their welfare best served by access to growing world markets (where their price must be competitive) and by measures that allow them to expand acreage or intensify the use of presently cropped lands to serve those markets. Past conservation policy was based on compensating farmers for production restraint, but it acted in a depressed market. In the rising market we project, production restraint would be both unpopular and costly. Public compensation in return for production curbs likely would have to be prohibitively expensive to be effective.

If our soil and water resources are threatened sufficiently by the accommodation of high demand, then measures to restrict demand could be one response. Even though we do not think that severe measures of this sort are justified at present, it is useful to consider what might be done.

The most obvious candidate is exports. The United States exports over one-third of its grain and supplies about 60 percent of world trade in grains. No one can accurately foresee the future, but if the United States maintains its share of world markets, as our projections assume, then most of the large domestic increase in production will go into exports. Restricting exports would reduce output and greatly ease the pressure on U.S. agricultural land.

Restricting exports would of course be very unpopular with domestic producers and counter to the U.S. policy of promoting freer trade. It also would antagonize buyers whose amity we cherish. In a hungry world, restricting food exports might be considered an inhumane policy. It also would be argued that we need the exchange generated by farm exports to finance petroleum imports. In response, it could be argued that the degradation of renewable resources--land and water--in order to satisfy a bloated and ephemeral demand for oil is a bad trade and that energy conservation or domestic fuel alternatives should be sought instead. Restrictions on exports could be managed so as to accommodate favored friends and customers, though not without damage to the world's trading system. Humanitarian responsibilities are not unlimited, and, in a world of national sovereignties, every nation must look after the health of its own renewable resources, since no one else can. Yet restrictions on trade surely do not deserve early attention as a conservation policy. Many other policies to control erosion that are consistent with high output should be explored first. If the problem becomes acute enough to warrant restrictions on trade, they could be instituted quickly.

Nonetheless, it seems likely that American agricultural products are sold too cheaply on world markets and, therefore, in greater quantity than might otherwise be the case. Both the subsidies that abound throughout U.S. agriculture and the environmental costs of production are not being reflected in price; therefore, the United States subsidizes foreign buyers, including both developed countries and East Bloc adversaries. The increased domestic environmental damage and soil degradation that result are real costs paid by current and future generations of Americans. In principle, there can be no objection to policies that correct for this underpricing or subsidy to consumers of agricultural products, especially of those foreign buyers who bear none of the domestic fiscal or environmental costs. The preferred solution would be to root out the subsidies at the source or to reflect all external costs in the price of agricultural commodities. However, that would be a major undertaking--the present structure of agricultural policies is simply too deeply embedded in our system. An export levy might be a more feasible way of attacking the foreign component of the problem, but it carries the disadvantage of penalizing those producers who may not use subsidies or create environmental problems. Politically, and perhaps constitutionally as well, it also would be difficult to accomplish.

Export demand also may be limited by fostering production abroad. To the extent that demand is from developing countries and our concern is humanitarian, there should be no objection to policies that assist them and spare our own soil resources. This would involve transferring and developing technology suitable for their conditions and assisting in the development of new land and water abroad. By creating the capacity for others to help themselves, we respond to the themes of self-reliance and independence now so strongly asserted in the developing world.

Domestic demand is not a dynamic force for agricultural expansion, but its composition could be altered so as to reduce pressure to produce by shifting the diet away from grainfed meat animals. To do so need not result in any loss of nutritional adequacy and might even produce health benefits. There is growing evidence that meat eating on our present scale is not necessary for a healthy diet and may even be harmful to it. The confirmation and publicizing of this evidence could significantly alter our diet and demand for grain. Higher prices for feed grains, as they become reflected in the price of meat, also serve to curb the domestic appetite for meat.

Finally, it should be possible to spare cropland by making better use of forage resources and unconventional feeds. Improved management of public rangelands is one obvious possibility. At the same time, much agricultural and forest biomass not previously used by animals is proving adaptable as feed.

Soil Loss and Social Responsibility

We concluded in the last chapter that the case for public intervention to protect the productivity of the land against erosion is not as strong as often supposed, but that there are circumstances when it has merit. A consensus may emerge that society's responsibilities to future generations require a measure of erosion control beyond what farmers will apply in their own interest. But is there any basis for making the cost of this additional effort a responsibility of current private holders when they impose no current off-site costs on others? The legal foundation for restricting the use of property is the need to protect public health and safety, but it stretches the police power severely if it is asked to regulate matters that are so remote in time and that affect no identifiable current victims. To the extent that such problems are dealt with at all, it is through an implicit intergenerational bargain involving the whole society rather than individual actors. For example, the current consumption of exhaustible resources that may deprive a future generation of their use is not something for which each present individual is held directly accountable. Rather we view broadly borne restraint and social investment in resource replacing technology as the way to meet our obligation to the future. Likewise, if we wish to protect the rights of a future generation in a productive soil, the equitable way of meeting the current costs of doing so would be to make them broadly bórne.

The farmer has both operational control and constitutionally protected property rights on his land. So long as damage is confined to his own acres, it is not clear that he can legally be compelled to correct it. Further, it might appear inequitable for society to require that its interest in preserving the soil be accomplished at his particular expense rather than being broadly borne. This principle is even more important because of the farmer's operational control over the land and the fact that he must be the instrument for effecting whatever measures are taken.

Soil Conservation Policies

Our review of current policies suggested much room for improvement. Yet, the accomplishments of the present system should not be denigrated. Farmers are well informed on what can be done to limit soil loss. They appear to have a decent appreciation of what conservation steps are in their own best interests and have gone far toward applying them. Allegiance to a conservation ethic is widespread; under its influence many farmers undoubtedly undertake conservation measures not justified by private economic calculations. In effect, they pay a self-assessed tax toward a public goal. Other farmers have accepted contracts and installed conservation practices on a cost shared basis, thereby helping to restrain soil loss. In this case they may also have enhanced the value of their own property at public expense--an outcome that can result from any public subsidy to private action.

Under a low or moderate demand scenario, current policies might suffice to achieve a socially acceptable level of soil loss, especially if they were targeted better on key areas. However, the demand scenario we have projected carries the risk of soil losses far higher than at present. The public is not likely to accept the 84 percent increase in soil lost from cropland or average per acre losses of over 7 tons per year from the expanded area in crops that appear in our run 2 (see table 7-2).

Whether or not the pressure becomes this acute, existing programs can be better focused. The GAO study comments are especially applicable here (GAO, 1977). Detailed farm conservation plans are not needed in many circumstances, yet SCS spends an inordinate amount of time on them to the neglect of more critical problems. The passive approach of SCS means that scarce manpower resources go to those requesting help rather than to those who need it. Meanwhile, because ASCS allotment programs are based on historical acreage, they encourage farmers to plant crops on land that should be in other use. The GAO report recommended that SCS systematically seek out farmers with the most critical erosion problems, tailor its conservation plans to solve those problems without being overly elaborate in the plans prepared, and coordinate with other agencies to relieve program conflicts. Likewise, the GAO criticized the lack of priorities in assigning ACP money by area and form.

Not all of the GAO criticism seems well directed, however. While, from a social standpoint, the government should not be paying for measures that the farmer would find profitable to increase his productive capacity, why does increasing capacity differ from preserving capacity for which we have long accepted a government role? Perhaps the criterion for government support in either case ought to be whether it is profitable to the farmer. If not, then we should aim at supporting those lands and measures that offer the highest favorable social benefit/cost ratio for public investments at whatever discount rate, if any, is selected. In particular, short-term measures that are highly cost effective in controlling erosion should not be dismissed in favor of longer term structural measures that are less so.

The present system has had only moderate success in dealing with existing pressures on the land. It contains a social element in appealing for the practice of good husbandry, but mostly it has relied on the farmer acting in his own interest and has provided him with information and technical support to do so. In addition, it has included subsidy elements in the form of cost sharing for installing practices that are not justified by the farmer's private interest. Fine tuning of this system would be a help, but when commodity prices make it attractive to abandon conservation measures, it is difficult to resist the pressure. The experience with the Great Plains program cited earlier is the best evidence of this, and elsewhere the widespread return to crops of acreage once idled by conservation programs is additional evidence. If a high export scenario prevails, more compelling inducements to maintain conservation practices will be required.

The draft RCA study establishes targets (essentially the achievement and maintenance of "T" values) and lists alternative strategies for achieving them (USDA, RCA II, 1980). Since the export projections used in that study are lower than ours, their policy recommendations are directed at a less acute problem. Thus, while their alternatives of organizational reform and reassigning responsibilities to make current programs more effective could help, the crunch comes in the choice of devices to induce compliance.

Inducements may be either positive or negative--i.e. desired behavior may be rewarded or undesired behavior penalized. One proposal that has received much attention is the idea of cross compliance (Benbrook, 1979). In negative form, this would penalize farmers who fail to employ good conservation practice by withdrawing from them the benefits of other federal programs, such as price supports. In positive form, it would offer favored treatment in benefits from these programs to farmers using good practices. Farmers prefer the second approach and have been highly critical of the penalty version. While that preference is understandable, it is hard to see why farmers should be granted a vested right to public benefits when they ignore public policy.

Yet the idea of cross compliance has several other drawbacks. Not all farmers use the benefit programs, so the reach of this device is incomplete. It is also unselective and would not focus on those lands most in need of attention. Moreover, in a high output scenario, federal price support programs would become of less interest to farmers at the very time when erosion control becomes a more acute problem. Finally, from the standpoint of social strategy, it may be unwise to tie a long-term concern with soil erosion to an array of agricultural policies that may not deserve continued support on their own merits. To do so would tend to embed those policies further into the system at the very time when they otherwise might be modified.

Taxation can be used either as a reward or penalty. Property taxes, for example, could be lowered for those following good practices. This would require state/federal cooperative arrangements to encourage states to participate. But manipulation of general purpose taxes for special purposes, while widespread, should not be encouraged. A punitive tax on socially undesired behavior could be quite effective in theory and would have certain efficiency benefits. Thus, a soil loss tax would provide powerful incentives for the farmer to employ conservation measures while allowing him full latitude to do so in the next most efficient way (Seitz and coauthors, 1979). It is objected that such a tax would be difficult to administer, though it is hard to see why it should be more so than a cross compliance scheme. The tax could be made operative only when losses exceed the "T" level. Note that a soil loss tax, if unrelated to sediment delivery beyond the farm, places the burden on the farmer to achieve the social objectives of preserving his land's productive capacity. Therefore, while the tax may be employed as an incentive, it could logically be combined with subsidy programs that relieve the burden on those willing to cooperate.

Present subsidies for conservation practices are in the form of cost sharing programs. Being voluntary, they do not attract all farmers and, as we have seen, not necessarily those with the worst problems. Since the farmer still must pay some part of the cost, he has little incentive to act unless the measure increases his capacity in proportion to his own investment. Further, he is tempted to install those practices for which money is available, even though these may not be the most efficient ones in his case. As a consequence, the public gets poor value for its money.

A system of contracts between farmers and government to reduce soil loss has much appeal. It would allow the program to target on those areas and on the specific acres that most deserve attention. The contract could be written to specify performance, allowing the operator to arrive at the most efficient means of achieving it. Good cultivation would not be at a disadvantage in this system; thus the public not only would get what it pays for but would not pay more than necessary. As the RCA study points out, the contracts would be flexible, with provisions for the owner to "buy out" if market conditions change, and he could trade off other conservation measures for those he wishes to abandon if the government agrees (USDA, RCA II, 1980). A possible weakness is that the farmer might be able to exact payments on land that he would not crop for his own reasons, but that defect has been present in past conservation programs. Administrative costs are expected to be high.

The Great Plains Conservation Program operates on a contract basis but, as was noted by the GAO, it has not enlisted the desired participation rates or been able to restrain the reversion of grassland to crops at the end of contract periods. Evidently contract prices must be realistic in meeting the cost to the farmer of doing society's work for it, yet they should not exceed the value of the social benefit derived. If the farmer is fully compensated for all costs, including opportunity costs, there is no reason for him to resist the contract, and the same applies to renewal if the contract reflects the changed conditions prevailing at that time.

Nonetheless, the program need not be entirely voluntary. Where reasonable contracts are offered, the operator's defense for declining to take conservation measures vanishes. If he still resists, he could fairly be coerced, either by tax or regulatory measures. A soil loss tax paid by those who decline to contract would encourage participation. It seems preferable to regulation, for it gives the operator yet one more chance to do the job in his own way while still protecting the public interest.

The threat of regulation to encourage farmers to contract is consistent with protection of the public interest. Major reliance on regulation alone is apt to disappoint, however, for it is very difficult to compel a farmer to undertake specific measures on his own land. Moreover, if the purpose is primarily conservation beyond that which is profitable to the farmer, then the equity of imposing the costs on him is questionable when damage is confined to his own land. Thus, regulation appears better suited as a backstop to a system of contracts, but it would appear inferior in that role to a soil loss tax.

Any subsidy scheme to induce good soil conservation is premised on acknowledgement of the sanctity of an owner's property rights. In effect, it concedes his right to use his property in a socially undesirable way unless he is offered compensation to refrain. This is a familiar but not unchallenged position. Where property is not involved, we do not allow unconstrained pursuit of private purposes without regard to social welfare. It is troublesome to contemplate a society where antisocial behavior is restrained by a complex system of compensating payments in perpetuity. Legal sanctions are justified when harm to others is clearly demonstrated, as with off-farm damages of erosion. But is society's preference (or obligation?) to maintain the resource base intact sufficient justification to override the legal right of an individual to dissipate his capital (soil) in favor of current returns? This entire dilemma could be avoided by acquisition of certain property rights, but if we want to "take" some aspect of property rights, then it should be done cleanly, openly, and once for all.

A contract system does not by itself ensure that an optimum amount of erosion control is purchased by public funds. In some cases the cost of soil conservation may be so high that even from a social accounting standpoint it is rational to mine the soil rather than preserve it. Before conceding that to the operator, however, he should be made to face the off-site consequences of such action.

Water Pollution Control Policies

The preceding section has dealt with policy to control soil loss so as to maintain productive capacity. Such policy confronts a landowner who can argue that his property is his to manage as he sees fit so long as he does not currently harm others; a legitimate public interest here must accommodate equally legitimate property rights. But if the owner's actions do harm others, then it is usually accepted that he is accountable and may be restrained by appropriate measures. Farm-related water pollution fits the latter case. Since much farm-related water pollution is entrained with sediment, the two problems of erosion and water quality often are treated as one, though the distinction can be important in designing program emphasis or in assigning cost incidence.

Throughout the private sector, measures for environmental amelioration have been made the responsibility of the offending party. Auto manufacturers must design cars to meet standards at their own expense, utili-
ties pay to control emissions at standard levels, and best available control technology is required of manufacturers discharging to streams. The
final incidence of these costs depends on market structure, but the costs
do generally become reflected in prices paid for the products and, therefore, affect production and consumption decisions.

Nonpoint sources other than agriculture are subject to a variety of
controls, but the agricultural sector generally has been accorded more
lenient treatment. It is hard to understand on what principle this should
be so. Sometimes it is argued that the costs of such measures are difficult to pass on to consumers because markets are competitive and therefore
costs come to rest on producers. What kind of an argument is this? If
costs were promptly reflected in decisions on how much, what, and how to
produce and ultimately in commodity prices, this would help to reduce
farm-related pollution.

Administratively and politically, however, it is very difficult to
impose both the cost and the responsibility for control on the farming
sector. The number of units to be controlled is large, data on effluents
from a given farm are lacking, and the consequences of their discharge
hard to establish. Moreover, EPA has no bureaucracy in the countryside
and would not be welcomed there, and USDA, which does have what are called
"delivery systems" is loath to be perceived as regulator rather than ser-
vant of its clientele. Existing programs for farm-related water pollution
control are mostly based on education and voluntary measures, with the
latter sometimes induced by cost sharing. Indeed, the cooperative and
voluntary tradition employed by institutions in touch with farmers is so
much a part of farmer's expectations that abrupt change seems quite infea-
sible. Selective regulation can be admitted--in some states where problems are acute local authorities can act through regulation--but the enforcement of generally applicable rules is much harder to envisage.

Projects under the Rural Clean Water Program would represent a concentrated attack on some of the more acutely affected areas, albeit with
the usual congressional sensitivity to regional distribution. Since one

can have reservations about the equity and the economic consequences of public subsidy to the abatement of privately generated pollution, it might be argued that this program should be kept small. In any case, a limited budget should help to focus the program on the most troubled areas where the returns from public investment should be high. Given the rather poor documentation of farm water problems and the damages they cause, this seems to be a politically realistic approach.

Even within this basic framework, state, and local authorities, and Soil Conservation Districts could be more aggressive in disciplining blatant offenders. Direct counselling and local enforcement seems to have had some effect in California, and there is no need to await ponderous federal legal processes before attempting these simpler approaches.

Another possibility within the existing framework is to make farm conservation plans more explicitly water quality plans as well. The plans already are criticized as overly elaborate, but most of them were prepared with an eye to soil conservation rather than water quality, and they may have neglected simple and effective measures conducive to quality. As new plans are prepared or old ones modified, this defect should be repaired. Experience with RCWP and with some experimental model implementation projects (MIPs) should give SCS an opportunity to sharpen its approach to water quality control.

If our projected scenario prevails, however, then water quality problems attributable to agriculture likely will be much more severe than at present. They will be mostly sediment related and will increase in all regions, but particularly in the nation's agricultural heartland--the Corn Belt, Southern and Northern Plains, Mississippi Delta, Appalachia, and the Southeast (see table 7-3). There will be wide variations among regions, however. Measured by the increase in amount of sediment delivered the Southern Plains will be by far the most seriously affected. This is a region now little concerned about or prepared to cope with increased pressure. Among the major producing regions the Corn Belt will be least affected. The Northeast, Mountain, and Pacific regions will all experience substantial percentage increases in sediment delivered, but the absolute amounts will remain relatively small. These disparate trends, while stressing the need for programs that are responsive to local conditions, also suggest that it may be necessary to stiffen local resolve in some cases.

Whenever federal programs differentially affect some areas there is a cry of unfairness and a call for equalizing payments. If there is unfairness, it is in the needless application of uniform national point source regulations that cause differential effects. However, water quality control is a local responsibility. Except for the broad goal of fishable and swimmable waters, nonpoint pollution control will be pursued in relation to state established stream standards. There are no effluent standards for nonpoint sources. Differential regional impacts could occur through simple nonenforcement or through competitive degradation of stream standards in order to retain local production. If stream standards are set in some objective way and are enforced, then any depressing effect that water quality control may have on local farming activities is a reflection of the comparative advantage of local land resources and need not be equalized by national payments. Some areas may be disadvantaged, and local farmers may lose income and capital values, just as any segment of society may do by change, but they are not unlawfully or unreasonably deprived according to any standard of equity.

It is clear that under our projected scenario EPA may need to take a sterner view of 208 plans and, especially, of their implementation in some districts. States could be encouraged to identify quality problems more precisely and to coordinate with federal programs so that water quality measures could more effectively go piggy-back on conservation measures where conservation receives high priority, or so that quality can receive direct attention if conservation does not.

The affected states will have the responsibility to implement stronger measures for farm water quality control. They cannot expect equalizing federal grants to protect their agriculture from adjustment to the need to protect quality. However, federal money is likely to be available for measures primarily aimed at conservation, and fairly inexpensive adaptations to such measures could have important quality implications. Thus, states that are threatened by severe quality problems would be advised to strongly support conservation programs for which federal money is available and to backstop such programs with provisions for penalizing those farmers who fail to cooperate. If the contract/soil loss tax combination is available, it should provide a state with such an opportunity. However, the water quality dimensions of conservation plans must be developed in cooperation with the federal technical services.

It is worrisome to note the asymmetrical character of conservation programs, which we have argued should be federally supported to the extent that they protect a social rather than a private interest, and water quality programs that, being an external cost, should be the farmer's responsibility and become reflected in price and production decisions. The implications of these two positions for the location and profitability of production are quite different. Were there no vested rights to consider, conservation could be required up to a socially established level just like quality control, with the operator absorbing any loss of income and capital that might imply. But in our view equity argues against that. Our position on responsibility for environmental protection is consistent with practice elsewhere in the society. However, if we choose inconsistency with other environmental rules and are willing to ignore the budgetary consequences, we can subsidize quality control just as we do conservation.

In practice the two principles can support one another. A soil conservation program makes sense from a social standpoint on much private land. It may require the retirement of some land from crops and the less profitable use of other land. The farmer must be given inducement to do that which is socially rational. A realistic subsidy is at the center of this. But his sense of responsibility can be accentuated by making him responsible for the off-site consequences of soil loss. The conservation objective pursued by a performance contract or like device can be backstopped by a soil loss or sediment tax grounded in quality control for those unwilling to accept the conservation objective freely. Soil miners at least would be deterred to the extent of paying for off-site damages. And in all cases the incentive to use the most efficient methods for obtaining target objectives would be preserved. Moreover, both subsidy and tax could be, and should be, varied by area to reflect the real social value of the objective sought.

Research as Policy

Increased environmental stress from agriculture originates both in the expanded scale of production and in the technologies employed. New technologies often are seen as posing added threats, but if they are pursued with the need in mind to accommodate production while diminishing environmental impacts, then the outcome can be quite different. Moreover, if we are to avoid both surprises and needless fears, there is a need for better information on the consequences of changes in technology and scale of production. Thus, research on which to base both control actions and environmentally benign new technologies can be an important policy tool. The principal policy issues are those of the scale and the focus of the research.

Concern about research has been implicit in much of the previous discussion in this volume. We noted that knowledge of where to focus programs for environmental amelioration is skimpy. State 208 plans, intended above all to identify nonpoint sources of water pollution, have not increased the precision of this information as much as was hoped. EPA could help by identifying and publicizing the more successful problem identification approaches found in state plans.

There is always a temptation for a single mission regulator to attack those problems that he perceives directly without full attention to the possibly adverse displacements that this may cause elsewhere for other objectives or, indeed, even for the one intended. Although EPA has the principal environmental regulatory authority, it does not have detailed knowledge of agriculture that would permit it to assess these tradeoffs, and it has little incentive to research them thoroughly or to seek technologies that may restructure them. One course might be for EPA to give firmer long-range guidance on ambient needs and then allow the states and/or USDA to structure controls and related research programs to meet those needs.

However it is triggered, detailed research is needed on the most effective ways of meeting ambient goals. Several possible directions for research come to mind. Extension to other crops of the principle of integrated pest management already showing results in cotton deserves more attention, as does conservation tillage. The use of fertilizer also needs study. While the leaching of nitrogen presently is a spotty problem, our projections suggest it may become more severe and more general.

Environmental concerns need to be a more active component of agricultural research aiming at yield increasing or cost reducing technologies, for if the latter are not environmentally tolerable, they will fail. Increasing yields, especially on the best and least vulnerable lands, can relieve pressure on the environment as well as restrain economic costs. The rapid strides in the biological sciences of recent years so far have not been translated into new varieties on any great scale; there is an uncertain but possibly large potential to do so. Targets might include plants that are more efficient synthesizers, more pest or drought resistant varieties, and varieties with greater protein content. Conventional plant breeding and cultivation research might give attention to soybeans-- an important crop whose rate of yield increase is not impressive. Advances in these areas should permit higher yields without greater environmental hazard. A related possibility is the development of more unconventional feeds for animals using resources presently unutilized. Even the development of new markets can be an important innovation if they allow more intensive (namely, corn-soybean combinations) use of good lands or the substitution of higher yielding crops (e.g., sunflowers for wheat).

It would be helpful for agencies to be more aware of the productivity and research implications of regulatory actions. Regulation that restricts the use of current technology (e.g., pesticide use) may divert scarce research resources into compensating for the lost techniques. And regulation may discourage private spending on research (e.g., in pesticides or genetic engineering) if it is not sensitively designed.

Agricultural research is especially susceptible to policy because so much of it is the province of government. Agricultural commodities are undifferentiated and are sold by small firms that are unable to undertake costly scientific research of the kind needed. Research by private vendors of supplies and services can be an important contribution in some areas, but the government has an uncommon opportunity in this field to set the priorities and the scale of the program in response to an integrated appraisal of social needs.

A Final Word

American agriculture has always been one of our nation's greatest
strengths. From the first Thanksgiving to the factory farm, it has responded to the needs placed upon it by our society. The challenge posed
in this study is whether agriculture can meet the high level export demands that it may face in the future without irreparable damage to the
productive base or unacceptable cost to environmental quality. On the
whole, we conclude that it probably can, but that social controls are
likely to intrude more into the countryside than in the past if major
objectives are to be met. If we wish to avoid land use controls, then
pressures on the land can only be relieved by reducing demand or finding
land substitutes. If we want to pursue all objectives simultaneously,
then conservation and rural water quality programs must be operated more
efficiently and in a mutually supportive manner.

Any set of distant projections, such as those attempted here, is
subject to wide error. If the demand projections are too exuberant, much
of the potential problem vanishes. Others have been more conservative
than we in projecting demand, but we think that our scenario has sufficient plausibility to deserve attention. The implications of such a level
of production for the use of land and other resources is also subject to
interpretation. Per acre yields must rise if demand is to be satisfied
within the land resources that we are likely to be able to devote to
crops. We think that condition can be met, but we do not expect the past
yield trend to be matched. Fertilizer use will grow, but if better use
is made of that applied, it will not outpace production growth. Future
pesticide use will depend on the location of production and on the techniques of pest management and cultivation employed, but declining use of
insecticides on cotton and moderation of use on corn suggest that no ex-
plosion will occur in this area.

The primary consequence of expanded production that has held our
attention is damage to the soil resource from erosion and the consequent
damage to water quality as this is discharged as sediment to streams.
Thus among emerging environmental problems erosion control seems to deserve the highest priority. But erosion is not only a function of the

scale of production but also of the cultivation practices and the location of production as well. The spread of conservation practices, especially conservation tillage, gives promise that any given output can be produced with less damage to the soil. However, not all soils or crops are suited to this practice, and we have been more conservative in projecting its adoption than some other projections have been. Moreover, conservation tillage implies increased pesticide use, especially herbicides, with consequences that cannot be entirely foreseen; therefore other pest control techniques must continuously be explored. As a consequence of this (together with our demand projections), we foresee potentially grave problems from agriculture. They will not be moderated much by geographical production shifts, for any such moves will be toward vulnerable areas of the country.

We are confident that the nation has the physical capacity to meet agricultural demands and that the environmental consequences of doing so are on a scale that is manageable in principle. However, the latter will not occur automatically. There must be recognition that a more heavily stressed agricultural system may demand closer control if it is to continue to expand without damage to productive capacity and environmental quality. Much better monitoring and data on farm-related pollution should be sought so that adverse trends can be spotted early and policy designed to meet them. With these significant caveats, we believe that American agriculture will continue to meet its domestic and international challenges.

References

Benbrook, Charles. 1979. "Integrating Soil Conservation and Community Programs: A Policy Proposal," Journal of Soil and Water Conservation (July-August).

The Clean Water Act (Public Law 92-500).

General Accounting Office, Comptroller General of the United States. 1977. "To Protect Tomorrow's Food Supply, Soil Conservation Needs Priority Attention" (Washington, D.C., GAO, February).

Haith, Douglas A. and Raymond C. Loehr (eds.). 1978. "Effectiveness of Soil and Water Conservation Practices for Pollution Control," Draft manuscript (Ithaca, N.Y., Cornell University).

Harris, Louis, and Associates, Inc. 1980. "Survey of the Public's Attitudes Toward Soil, Water, and Renewable Resources Conservation Policy," January 17.

Holmes, B. H. 1979. Institutional Bases for Control of Nonpoint Source Pollution (Washington, D.C., Environmental Protection Agency).

Iowa State University College of Agriculture. 1978. A Technical Assessment of Nonpoint Pollution in Iowa. Mimeo. Assembled by Lucy Harman and E. R. Duncan (Ames, Iowa, March).

John Muir Institute. 1979. Erosion and Sediment in California Watersheds: A Study of Institutional Controls (Napa, Calif., John Muir Institute, June).

Loehr, Raymond C., Douglas A. Haith, Michael F. Walter, Colleen S. Martin (eds.). 1979. Best Management Practices for Agriculture and Silviculture, Proceedings of the 1978 Cornell Agricultural Waste Management Conference (Ann Arbor, Mich., Ann Arbor Science Publishers, Inc.).

Nagadevara, Vishnuprasad S.S., Earl O. Heady, Kenneth J. Nicol. 1975. Implications of Application of Soil Conservancy and Environmental Regulations in Iowa Within a National Framework, Card Report #57 (Ames, Iowa State University, June).

National Research Council (NRC), Committee on Prototype Explicit Analyses for Pesticides. 1980. Regulating Pesticides (Washington, D.C., National Academy of Sciences).

Seitz, W. D., D. M. Gardner, S. K. Gove, K. L. Gunterman, J. P. Karr, R. C. F. Spitz, E. R. Swanson, C. R. Taylor, D. L. Uchtman, J. C. Vaness. 1979. Alternative Policies for Controlling Nonpoint Agricultural Sources of Water Pollution, EAP-60015-78-005 (Athens, Ga., Environmental Research Laboratory, EPA, April).

U.S. Department of Agriculture (USDA). "Basic Statistics 1977, National Resource Inventory (NRI)," Mimeo revised (Washington, D.C., USDA, February).

_____. 1974. Our Land and Water Resource (Washington, D.C., USDA, May).

_____. 1980. 1980 Rural Clean Water Program (RCWP) (Washington, D.C., USDA, March).

____. 1980. "Summary, Parts I, II, and Environmental Impact Statement," Soil and Water Resources Conservation Act (RCA), Review draft (Washington, D.C., USDA).

____. 1980. National Summary Evaluation of the Agricultural Conservation Program - Phase I (Washington, D.C., USDA-Agricultural Stabilization and Conservation Service).

Wischmeier, Walter H., and Dwight D. Smith. 1965. "Predicting Rainfall-Erosion Losses from Cropland East of the Rocky Mountains," USDA-ARS Agricultural Handbook, no. 282

Young, Keith K. 1980. "The Impact of Erosion on the Productivity of Soils in the United States," in M. DeBoodt and D. Gabriels (eds.) Assessment of Erosion (New York, John Wiley and Sons).

APPENDIXES

Appendix to Chapter 2

DERIVATION OF PROJECTIONS OF PRODUCTION

The projections for grains and soybeans were made in three steps:
(1) project growth in world trade in each commodity; (2) project the U.S.
percentage of trade; (3) project domestic use. The projections for cotton
were derived from a USDA study (Collins and coauthors, 1979) described
below.

Projections of World Trade

Wheat and Coarse Grains[1]

The projections of world trade in these commodities are based on
analysis of trends in production and consumption in importing countries
and groups of countries. These are the importing developing countries,
Western Europe, Eastern Europe, the Soviet Union, the People's Republic
of China (PRC) and Japan.

Importing Developing Countries.[2] Much of the increase in demand for
food in the less developed countries (LDCs) has come from the relatively
few among them which have higher than average income and income elasticities of demand. This is an argument for considering these countries apart
from the other LDCs. We decided that for our purposes the improvement in
our projections from doing this would not justify the extra effort required.

Projections of food consumption typically are based on projections of
population, per capita income, and income elasticities of demand. The

---

[1] World trade data are for coarse grains rather than feedgrains as
defined in table 2-1. Coarse grains as defined by the USDA are feedgrains plus millet and "mixed grains".

[2] For wheat these are all countries of Asia, Africa and Latin America
except the PRC, Japan, South Africa, and Argentina. For feedgrains these
countries, as well as Brazil and Thailand, are excluded.

World Bank (1980, pages 142-143) projects population growth in the importing developing countries as defined here at 2.2 percent from 1980 to 2000.
Although the Bank report does not indicate this, rates of population
growth in these countries are declining, and one recent projection is that
by 1995-2000 rates will be 1.62 to 2.00 percent annually.[3] Accordingly,
we assume that in the countries of interest to us population growth between 2000 and 2010 will average 1.7 percent per year. In this case,
average annual growth over the full period 1980 to 2010 would be 2.0 percent.

The World Bank (1980, page 99) shows that in the 1970s real per
capita GNP in the developing countries increased 2.8 percent annually,
down from 3.1 percent in the 1960s. The Bank expects per capita GNP in
these countries to grow only 2.6 percent from 1980 to 1985 because of the
steep increase in oil prices in 1979. Between 1985 and 1990, however, the
Bank expects GNP per capita in these countries to grow 3.3 percent annually, reflecting a successful transition to a world of permanently higher,
and continually increasing, prices of oil. We assume that growth at 3.3
percent annually will continue to 2010. In this case, real per capita
GNP over the entire period 1980-2010 will average 3.2 percent per year.[4]

The income elasticity of demand for grains in the developing countries is between .3 and .5, being higher for wheat and lower for coarse
grains for direct consumption (USDA, 1974, page 77). However, in countries with incomes sufficient to support rising demand for meat, the
income elasticity of demand for coarse grains for livestock is relatively
high. If real per capita income in the developing countries increases by
3.2 percent annually, the demand for animal products, and hence for coarse
grains, is likely also to grow steadily.

In summary, if population in the developing countries grows 2.0 percent annually from the late 1970s to 2010 and per capita income grows 3.2
percent, then demand for grains in those countries could reasonably be
expected to grow 3.0-3.5 percent per year over the period as a whole.

---

[3] Projections by the Community and Family Study Center, World Bank,
United Nations, and U.S. Bureau of the Census, as reported in Tsui and
Bogue (1978).

[4] Schnittker Associates (1979) and the Food and Agriculture Organization (1979) project per capita income growth in the developing countries
at 3.2 percent annually from the mid-1970s to 2000.

Appendix Table 2-1. Average Annual Growth Rates of Production and
Consumption in Importing Developing Countries, Wheat and Coarse
Grains

	1966-1979 (%)	1972-1979 (%)
Wheat		
Production	4.9	3.5
Consumption	4.8	3.9
Coarse grains		
Production	1.0	3.4
Consumption	3.1[a]	4.3

Note: Calculated from the logarithmic time trends. Sources are
various issues of U.S. Department of Agriculture, Foreign Agriculture
Circular Grains. For wheat, the data are world totals of production and
consumption less Western Europe, Eastern Europe, USSR, PRC, Japan, United
States, Canada, Australia, South Africa, and Argentina. For coarse
grains, production and consumption in the above countries, plus Brazil
and Thailand, are subtracted from world totals. Coarse grains are corn
and sorghum for grain, barley, oats, rye, millet and mixed grains.

[a]1968-1979.

Appendix table 2-1 shows production and consumption of wheat and
coarse grains in the importing developing countries in two periods since
the mid-1960s. The periods were selected to reflect the impact of the
Green Revolution in these countries on production and of the increase in
oil prices on both consumption and production. With respect to wheat,
the most notable feature of appendix table 2-1 is the slowdown after 1972
in growth of both production and consumption. The slower growth of production likely reflects the relatively slower growth of the Green Revolution after the initial burst in the mid-1960s. The heavy dependence of
the Green Revolution technology on irrigation and the difficulty in rapidly expanding the irrigated area in these countries probably is a major
reason for the slower spread of the Green Revolution since the early
1970s. Higher prices and occasional disruptions in supply of nitrogen

fertilizer in this period may also have contributed to slower growth in wheat production.

The slower growth in wheat consumption after 1972 likely reflects the slowdown in growth of production. While there is no necessary connection between the growth of consumption and production of wheat in these countries, balance of payments constraints imposed by sharply higher oil prices after 1973 likely would have prevented wheat consumption from growing at the pre-1972 rate, given slower production growth. In addition, real wheat prices may have been higher after 1972, although we have not pursued this possibility because of the complexity of internal pricing policies in these countries.

We believe that the tendency toward slower growth of wheat production and consumption will continue from 1979 to 1985 and from 1985 to 2010. Consumption will grow more slowly because of lower population and per capita income growth. Slower production growth also seems likely, at least for another decade, because of increasing real prices of energy and the expense of extending irrigation. We think it not unlikely that by the 1990s high-yielding varieties of wheat will be developed which are less dependent on irrigation, thus easing an important constraint on expansion of wheat production. This is not predictable, however, and we have not tried to take it into account in making our projections of wheat production in the developing countries.

There is, in fact, no firm basis for projecting specific rates of growth in wheat production and consumption in these countries. We assume that production will grow 3.3 percent annually from 1979/80 to 2010. We assume that consumption will grow yearly by 3.2 percent to 1985 and by 3.0 percent to 2010. The resulting projections of wheat production, consumption and imports are shown in appendix table 2-2.

Appendix table 2-1 shows that, unlike wheat, production and consumption of coarse grains grew more rapidly after 1972 than in the preceding 5 or 6 years. The behavior of coarse grain production reflects unusually low production in 1972 and 1973 because of bad weather. Production growth since 1972, therefore, cannot be used as a guide for projecting growth after 1979. The fact is that production of coarse grains in the importing developing countries has increased very sluggishly since at least the

Appendix Table 2-2. Production, Consumption, and Imports of Wheat and
   Coarse Grains in Importing Developing Countries

(millions metric tons)

Crop	1979/80	2010
Wheat		
Production	80.8	217
Consumption	133.1	332
Imports	52.1	115
Coarse grains		
Production	117.0	273
Consumption	139.9	399
Imports	22.9	126

Source: 1979/80 from USDA, Foreign Agriculture Circular Grains,
August 13, 1980. Projections for 2010 described in the text. Changes in
stocks assumed to be zero. For wheat the data include all countries of
Asia, Africa, and Latin America except the PRC, Japan, South Africa, and
Argentina. For coarse grains these countries plus Brazil and Thailand
are excluded.

early 1960s. For the period 1961-1979, production grew 1.3 percent annually. From 1974 to 1979, the annual rate was 1.4 percent.[5]

The acceleration in the growth of coarse grain consumption after 1972
reflects a rapid increase in consumption by the OPEC countries. These
countries consumed about 11.0 million metric tons of coarse grains in 1968
and about 11.5 million metric tons in 1972. Thereafter, their consumption
grew sharply, reaching 18.9 million metric tons in 1978.

While the growth of coarse grain consumption in the OPEC countries
was especially dramatic, consumption in developing countries not members
of OPEC increased 2.7 percent annually between 1968 and 1978, with even
faster growth--4.1 percent--after 1972.

The growth of coarse grain consumption in the developing countries is
consistent with their rising income and a consequent shift in diet toward
more animal protein. If per capita income in these countries grows as we
have projected it, then the proportion of animal protein in their diets
should continue to increase, with a consequent steady expansion in con-

----

[5]These and subsequent figures on production and consumption of coarse
grains are from various issues of USDA, Foreign Agriculture Circular Grains.

sumption of coarse grains. Consumption is not likely to continue to grow at the pace set since 1972, however, if population and per capita income growth in these countries slow down in accordance with our projections. We assume that from 1979/80 to 2010 coarse grain consumption will grow 3.5 percent annually. At this rate of growth, consumption would be 399 million tons in 2010.

We think it likely that this rate of growth in coarse grain consumption in the importing developing countries will stimulate coarse grain production to grow faster than in the past. Should production continue to grow at the sluggish pace experienced in 1961-1979 (1.3 percent annually), coarse grain imports in 2010 would have to be 220 million tons to satisfy the projected levels of consumption. Valued at 1979 FOB Gulf Ports prices of corn and grain sorghum, this volume of coarse grain imports would cost the importing developing countries approximately $26 billion. The comparable value of 1979/80 imports was about $2.7 billion. We argue below that these countries will be able to finance a substantial increase in their grain imports over the next several decades, and perhaps they would be able to manage an import bill for coarse grains of $27 billion. Long before the bill reached this magnitude, however, it is likely that these countries would seek ways to stimulate their own production of coarse grains so as to keep the bill to more manageable size. We think their search would be rewarded. Until the last few years, most of the effort of international and national food research institutions was devoted to development of higher yielding varieties of rice and wheat. Research on coarse grains was relatively neglected. Recently the balance has begun to shift toward coarse grains, with the Research Center for Improvement of Maize and Wheat (CIMMYT) in Mexico taking the lead. We expect that an increasing proportion of national and international research resources will be devoted to developing new technologies for coarse-grain production. While we cannot now point to any major breakthroughs, we believe the effort will soon begin to pay off in higher rates of coarse-grain production in the importing developing countries. Accordingly, we assume that production will grow 2 percent annually from 1979/80 to 1985, and that the yearly rate will average 3 percent from 1985 to 2010.

Appendix table 2-2 shows the projections of production, consumption and imports of wheat and coarse grains in the importing developing countries in 2010. Could these countries afford the projected levels of imports? We think they could. Valued at FOB Gulf Port prices of 1979, average wheat and coarse grain imports of these countries in 1979/80 cost about $11.3 billion. Valued at the same prices, our projections imply that the import bill for wheat and coarse grains would rise to $34 billion in 2010. In 1978, merchandise exports of the importing developing countries were $253 billion (World Bank,1980, pages 124-125). Their imports of wheat and feedgrains in 1979 prices ($11.3 billion) were 4.5 percent of 1978 exports. Exclusive of the OPEC countries, grain imports of importing developing countries in 1979/80 were about 7 percent of their merchandise exports in 1978.

Our projections of growth in population and per capita income in the importing developing countries imply that their real GNP will increase 5.0-5.5 percent annually. Export growth likely will keep pace, or approximately so. Our projections of grain imports of these countries, valued in 1979 prices, indicate annual growth of 3.7 percent. Thus, the value of these imports would decline relative to the value of exports. Since these countries were able to finance the grain imports of 1979/80 (food aid being relatively minor), it is reasonable to assume that they should be able to finance them also in 2010 when the imports likely will be smaller relative to exports than in 1979/80. Of course, if real grain prices are higher in 2010 than in 1979, the financing problem would be more difficult, and debt service may lay a relatively higher claim on export earnings than in 1979. Moreover, the problem of financing imports will not be equally distributed, those countries with low export-grain import ratios having a more difficult time than countries with high ratios (e.g., OPEC). Nonetheless, we believe the importing developing countries as a group will be able and willing to manage the financial burden implied by our projections of grains.

Other Countries[6]

Projections of Wheat Imports. Per capita consumption of wheat in
Western Europe, Eastern Europe and Japan was constant in the 1970s, and
there was a small declining trend in imports, reflecting expansion of
wheat production relative to consumption in the European Community (EC).
We assume that these tendencies continue to 1985, that is, that wheat
consumption will grow only with population and that imports continue to
decline at the same rate as in the 1970s. In this case, wheat imports in
these countries would be 16 million metric tons in 1985, compared with 17
million in 1979/80 (USDA, August 13, 1980).

Our projection for 2010 assumes that trends in production, consumption
and imports established in the 1970s continue beyond 1985 to 2010. In that
case these countries would import 13 million metric tons of wheat in 2010.

In the 1970s, the Soviet Union shifted from being a net exporter of
wheat to being a net importer and over the decade imports fluctuated widely, from .5 million metric tons in 1970 to 16.0 million in 1980. The
average for 1970-1980 was 7.3 million tons. In the year which ended in
June 1981, the USSR imported 16.0 million metric tons of wheat, and the
USDA expected the figure for 1981/82 to be 19.0 million tons (USDA, December 15, 1981).

The behavior of wheat imports by the USSR reflected both year-to-year
fluctuations in the weather and a political decision by the Soviet authorities to improve the diet of the Russian people. This decision is not
likely to be reversed. It suggests that the USSR will continue to be a
net importer of wheat.

While the shift of the Soviet Union from being a net exporter of
wheat to being a net importer is clear, the annual fluctuations in imports
have been so marked that we have no clear basis for extrapolating the
trend. Accordingly, we arbitrarily assume that in 2010 the USSR will
import 12 million metric tons of wheat.

From the early 1960s through the mid-1970s, the PRC regularly imported
2 to 6 million metric tons of wheat per year (USDA, May 1976). Since the
mid-1970s, imports have risen, reaching 13.8 million tons in the year end-

_____

[6]Western Europe, Eastern Europe, USSR, Japan, and the PRC.

Appendix Table 2-3.   Wheat Imports

(millions metric tons)

	1979/80	2010
Western Europe, etc.	17.2	13
USSR	12.5	12
PRC	10.2	20
Total	39.9	45

ing mid-1981, with 13.5 million tons expected by the USDA in 1981/82 (USDA, December 15, 1981). This increase in wheat imports by the PRC evidently reflected a generally more open trade policy adopted by the country in the late 1970s. Given the formidable problems faced by the PRC government in feeding its own people adequately, we believe the tendency toward rising wheat imports will continue. We have no sound basis for projecting these imports to 2010, however, so we arbitrarily assume they will come to 20 million metric tons that year.

Projected imports of wheat by Western Europe, Eastern Europe, Japan, the USSR and the PRC in 2010 are shown in appendix table 2-3. Corresponding imports for 1979/80 also are shown for comparison.

Projections of Coarse Grain Imports. For Western Europe, Eastern Europe, the USSR and Japan these projections are differences between projections of production and consumption. For Western Europe production is is projected to increase in accord with the trend established in 1970-1980.[7] In this case production in 2010 would be 139 million metric tons. (Production in 1978-1980 averaged 92.6 million metric tons.)

Coarse grain production in the USSR is projected to increase in accord with the trend established in 1960-1980,[8] in which case it would be 179 million MT in 2010. (Production in 1978-1980 averaged 93.8 million MT.)

[7]The trend equation for this period is $Y=74.2 + 1.62T$ ($r=.73$) where $T_1 = 1970$. Y is millions of metric tons of coarse-grain production. The source is USDA, August 13, 1980.

[8]The trend equation is $Y=44.6 + 2.63T$ ($r=.83$) where $T_1 = 1960$ and Y = millions of metric tons of coarse-grain production. Sources are USDA (May 1976, and August 13, 1980).

For Eastern Europe, we assume that production of coarse grains reaches 65 million metric tons in 1985 and then shows no further increase. Production in this region increased about 1 million metric tons per year on average between 1971 and 1981 (USDA, December 15, 1981). But from 1975 to 1979, the annual increase slowed to about .5 million MT. We assume the slower growth reflects underlying economic and physical constraints affecting feedgrain production in Eastern Europe and that these constraints will persist, allowing only limited additional amounts of production.

The projections of consumption of feedgrains in Western Europe, Eastern Europe, the USSR, and Japan[9] are based on projections of feedgrain consumption per capita. Per capita consumption in all of these areas now is less than in the United States, the greatest difference being between Japan and the United States (.15 metric tons per capita per year in Japan and .67 metric tons in the United States) and the smallest difference being between the United States and Eastern Europe (.53 metric tons per capita per year).[10] In the United States between 1960 and 1978, per capita consumption of feedgrains fluctuated between .57 metric tons (in 1974) and .76 metric tons (in 1972), but there was no trend. The average for the period was .67 metric tons. In Western Europe, Eastern Europe, USSR, and Japan, however, per capita consumption was rising, primarily in response to rising demand for grain-fed animals for meat.

We assume that this trend will continue to 2010, with each region approaching the current U.S. level of per capita consumption of feedgrains. Accordingly, per capita consumption is set at .30 metric tons in Japan, .65 metric tons in the USSR, and .60 metric tons in Eastern Europe. For Western Europe we assume the figure will be .45 metric tons.

The projections of per capita consumption of coarse grains were multiplied by projections of population in 2010 to derive the projections of total consumption in each region or country. Appendix table 2-4 shows the population data.

---

[9]Japanese production of coarse grains is insignificant. We treat it as zero. Consequently, projected consumption equals imports for Japan.

[10]All estimates of per capita feedgrains consumption are for 1978.

Appendix Table 2-4.   Population in Western Europe, Eastern Europe, the
USSR, and Japan

(millions)

	1976	2010	Average Annual Growth (%) 1970-1976	Average Annual Growth (%) 1976-2010
Western Europe	343.9	381	.55	.3
Eastern Euope	129.2	148	.68	.4
USSR	256.7	304	.90	.5
Japan	112.8	138	1.30	.6

Source: Population in 1976 and growth rates 1970-1976 are from the World Bank, 1978 World Bank Atlas (Washington, D.C., 1978). The projected growth rates are ours. Amy Ong Taui and D. J. Bogue, in "Declining World Fertility: Trends, Causes, Implications," Population Bulletin vol. 3, no. 4 (October 1978), project growth rates for the developed countries generally at .45 to .60 in 1995-2000.

From the first half of the 1960s to the second half of the 1970s, consumption of coarse grains in the PRC increased a little more than 3 percent annually, and until 1978 imports were negligible. In 1978, 1979 and 1980, however, coarse grain imports averaged 2.5 million MT per year (USDA, May 1976, and August 13, 1980).

As noted in the discussion of wheat imports by the PRC, the country evidently adopted a more open trade policy in the late 1970s, and the increase in coarse grain imports after 1977 may be a reflection of this. Whether this portends further growth in imports of coarse grains by the PRC is unknown since Chinese policy in this regard is unpredictable. Given the size of the country, however, the potential for increased imports of coarse grains clearly is substantial if the authorities pursue a policy of upgrading the quality of tne Chinese diet. Although we do not necessarily subscribe to this view, we nonetheless believe some increase in PRC imports of coarse grains is likely. Without pretending that we have a solid base for doing so, we project these imports at 15 million MT in 2010.

Our projections of coarse grains for Western E·rope, Eastern Europe, Japan, the USSR and the PRC are shown in appendix table 2-5. Data for 1979/80 are shown for comparison.

Appendix Table 2-5. Production, Consumption and Imports of Coarse
    Grains, Selected Countries and Regions
    (millions MT)

	1979/80			2010		
	Prod.	Cons.	Imports	Prod.	Cons.	Imports
Western Europe	92.0	110.7	23.8	139	171	32
Eastern Europe	62.0	71.4	10.7	65	89	24
Japan	.5	19.0	18.4	0	41	41
USSR	92.7	106.9	14.3	179	198	19
PRC	77.0	79.6	2.6	np	np	15
Total	324.5	387.6	59.8			131

np: not projected

Oilmeal

We are interested in U.S. production and exports of soybeans. However, in world markets soybeans are competitive with other sources of oilmeal, such as oilpalm. In recognition of this USDA data on world production and trade of oilmeal products is given in metric tons of soybean meal
equivalent.[11] The ratio of oilmeal weight to weight of soybeans in the
U.S. regularly falls between .78 and .80. We have converted the USDA's
data on world trade in oilmeal products to a soybean basis by dividing
them by .79. Our procedure, then, for projecting U.S. production of soybeans is in three steps: (1) project world trade in oilmeal products converted to soybean equivalents; (2) project the U.S. share of this trade;
(3) project U.S. domestic use of soybeans.

World trade in oilmeal increased rapidly in the 1970s. In terms of
soybean meal, the annual rate of increase was 7.8 percent between 1972 and
1980 (USDA, November 1977, and December 1979).[12] Since 1973, that is, in
the period since the run-up in energy prices, world trade increased at an
average annual rate of 9.0 percent. It is unlikely that the oilmeal trade
could continue to expand very much longer at such fast rates, and we assume
that growth will slow considerably between 1979/80 and 1985, namely to 5.5

---

[11]These data appear monthly in Agricultural Outlook.

[12]The trade figure for 1980 is a projection by the USDA based on 1979
production and analysis of previous relationships between production in
one year and trade in the following year. In recent years these projections have understated the actual increase in trade.

percent per year. In this case the world oilmeal trade in 1985, expressed
in soybean weight, will be 72 million metric tons. We expect continued
growth in trade from 1985 to 2010, but have no good basis for projecting
a specific amount. The demand for oilmeals, particularly those with high
protein content such as soybean meal, is stimulated by high and rising
income because those are the conditions which encourage high-protein diets.
Hence the major consumers of high-protein oilmeals at the present time are
the high-income countries. We expect continued expansion of oilmeal consumption in those countries as the demand for meat continues to rise. If
the developing countries achieve per capita income growth rates on the
order of 3.0-3.5 percent annually, as we have projected, those countries
also likely will become important consumers of oilmeal by the end of the
century.

At present, none of the high-income countries except the United States
are significant producers of soybeans or other sources of oilmeal. We see
no reason to expect this to change over the balance of the century. The
French have experimented with soybean production, but the results do not
suggest that France will become a significant producer of soybeans. Nor
does this appear likely for any of the other countries of Europe. Discussions with persons in Europe and the United States knowledgeable about
European agriculture indicate a consensus on this issue.[13]

The prospect of rising demand for oilmeal in both developed and developing countries with production confined to a relatively small number of
countries implies increasing trade in oilmeal between 1985 and 2010. We
project this at 3 percent per year on the assumption that the trend toward
slower growth in trade which we projected for 1979/80 to 1985 will continue
from 1985 to 2010. The result is shown in appendix table 2-6.

U.S. Shares of World Trade in 2010

In the four years 1976-79 the United States accounted for 44 percent
of world exports of wheat and for 65 percent of world coarse grain exports.

---

[13]In a listing of soybean producers in 1976-1978, Romania and Yugo-
slavia are the only European countries to appear. Romania was the more
important with average annual production of about 200,000 tons. The USSR
produced in average of 600,000 tons in the three years (USDA, 1978, p. 133).

Appendix Table 2-6. World Oilmeal Trade
(millions metric tons, soybean equivalent)

1979/80	2010
53.9	151

Source: For 1979-80, U.S. Department of Agriculture, Agricultural
Outlook (Washington, D.C., April 1980) p. 40. Figure is average trade for
the two years divided by .79 to convert from soybean meal equivalent to
soybean equivalent.

Between 1972 and 1980, the U.S. share of world exports of oilmeal varied
from 46.0 percent in 1975 to 59.5 percent in 1974. In 1979 it was 53.4
percent, and just under 53 percent in 1980. For 1972-80 as a whole the
U.S. share averaged 51.5 percent.

We assume that for wheat and coarse grains U.S. shares of world trade
in 2010 will be the same as in 1976-79, and that for oilmeal the share
will be 50 percent. For all three sets of commodities these shares are
slightly less than in the early 1980s, but essentially we are assuming no
change in shares from levels of the last 5 years.

The Case for Constant Shares

We noted in chapter 1 that our analysis of trends in crop demand,
input prices and technology suggests that real crop prices in the U.S.
will rise over the next several decades. We also noted in chapter 2,
however, that comparison of our production projections with those from the
USDA's NIRAP model indicated that our projections were consistent with
real crop price increases of 20 to 30 percent. Since the NIRAP model incorporates prices as well as domestic and foreign price elasticities of
demand, we concluded that our assumption of constant shares of world trade
in grains and soybeans was plausible. Here we present a more general
argument for the plausibility of the assumption.

The argument is based primarily on recent performance and lack of
evidence for a major change in the competitive position of other exporters
relative to the United States. Even good production performance in the
developing importing countries and the PRC seems likely at most to hold
the growth of their imports of grains within manageable limits. We al-

ready have indicated our belief that Eastern Europe has little potential
for additional production of grains and that the Soviet Union's shift in
the 1970s from being a net exporter of wheat to being a net importer will
continue. Continuation of the trend in coarse grain production in the
USSR, according to our projections, would enable that country to hold
imports to modest growth, but would not generate an exportable surplus.

Since the mid-1970s, the EC has been a net exporter of wheat, and
while the Community has been and continues to be a net importer of coarse
grains, its import deficit in these commodities declined from an average
of 13.6 million MT in 1970/71 to 9.9 million MT in 1978/79. This performance could be used to argue that the EC will emerge as an important chal-
lenger of the United States in world grain markets. We find the argument
unconvincing, however. For both wheat and coarse grains, the performance
of the EC was powerfully stimulated by the Community's Common Agricultural
Policy (CAP). The high internal prices adopted under the CAP both restricted the growth of internal demand for grains and gave impulse to production. Continuation of the CAP probably would increase the EC's exports
of wheat, and we allowed for that implicitly in our projections of wheat
imports of Western Europe and other regions shown in appendix table 2-3.
In our projections of coarse grains, we dealt with Western Europe as a
whole rather than the EC. Our projections showed that Western Europe would
continue to be a net importer of coarse grains even if the CAP stays intact
and coarse grain production continues to grow at the rate set in the 1970s.

The expansion of wheat exports by the EC does not necessarily mean
that the U.S. share of the world wheat trade would be diminished. The
emergence of the EC as a net exporter of wheat in the 1970s was at the
expense of Canada, Australia, and the Soviet Union, not the United States.
We have not undertaken a detailed analysis of wheat production and costs
in these countries so we are unable to explain why they lost ground in the
world wheat trade relative to the United States and the EC. Wheat yields
in Canada and Australia are well below those in the United States, although yield is a quite imperfect measure of production costs. However,
there is nothing in the trend of wheat yields in Canada or Australia to
suggest that those countries may improve their positions relative to the
United States. As noted in our discussion of wheat projections, the shift

of the Soviet Union from wheat exporter to wheat importer apparently reflects a policy decision to upgrade the diet of the Soviet people, a decision unlikely to be reversed.

With respect to coarse grains, the declines in the EC's net imports
in the 1970s meant that other countries' exports, probably including those
of the United States, increased less than they would have otherwise. However, as in the case of wheat, U.S. exports of coarse grains were less
affected by developments in the EC than exports of other countries. Indeed, the United States share of world trade in coarse grains increased
from 40-45 percent early in the 1970s to 65 percent at the end of the
decade.

Thus, the reduction in the EC's net imports of coarse grains already
is reflected in the U.S. share of world trade in these crops. Continuation of the trend toward smaller net imports in the EC, therefore, in
itself would not threaten maintenance of the U.S. share. Acceleration of
the trend would be a threat, but we see no reason to expect this. All of
the increase in EC production of coarse grains in the 1970s was due to
rising yields. The amount of land in these crops was constant, and knowledgeable people in Europe believe there is little potential for bringing
in additional land. Yields of coarse grains in the Community were almost
15 percent less than in the United States in the 1970s and rose at a somewhat slower rate.

In short, there is nothing apparent in the performance of EC production and trade in coarse grains in the 1970s to suggest that the Community might threaten the present U.S. share of world trade in these crops.
A detailed study might reveal undercurrents that our admittedly superfi-
cial analysis has missed, but in the absence of any such evidence we conclude that the United States could reasonably be expected to retain its
65 percent of world trade in coarse grains.

The ability of the United States in the 1970s to maintain its position in the world oilmeal trade is impressive since in this period Brazil,
and more recently Argentina, emerged as major exporters of soybeans and
soybean products. At the beginning of the 1970s, Brazil's and Argentina's
shares of the trade in soybeans and products were negligible, but by 1977
their exports of these commodities were almost one-third of world trade

while the share of the United States had declined to two-thirds (USDA, January 1979, p. 10). The fact that the United States maintained its position in the oilmeal trade generally, even while losing ground to Brazil and Argentina in soybeans, indicates that trade in soybeans and products was expanding faster than trade in other oilseed products.

Since the mid-1970s, the relative positions of the United States and Brazil-Argentina in the world soybean trade appear to have stabilized. Should U.S. exports maintain the current relation to Brazilian and Argentine exports, the U.S. share of the world trade in oilmeal would begin to rise if trade in soybeans and products continues to increase more rapidly than trade in other oilseed products. Should the U.S. share of trade in soybeans and products remain constant, its share of total oilmeal trade also would remain constant, even if trade in soybeans and products increases no more rapidly than trade in other oilmeal products. Some combination of a slower rate of increase in soybean trade relative to trade in other oilmeal products and slower growth in Brazilian-Argentine exports of soybeans and products relative to those of the United States also would yield stability in the U.S. share of total oilmeal trade.

On the basis of these considerations, we believe that U.S. exports of soybeans and products could reasonably be expected to continue to take half of total world trade in oilmeal.

U.S. Domestic Consumption

Our projections for 2010 are found by multiplying projections of per capita consumption by projected population of 279.6 million.[14] Annual per capita consumption of wheat in the United States fluctuated between 85 and 113 kilos between 1963 and 1978. The average was 100 kilos and there was no trend. We project 100 kilos for 2010.

As indicated earlier, per capita consumption of feedgrains in the United States was more variable in this period, fluctuating around an average of .67 metric tons, but there was no trend. We project .67 tons for 2010.

---

[14]Found by interpolating between the Department of Commerce's Series E estimates for 2000 and 2025.

Per capita consumption of soybeans increased from 64.4 kilos in 1963/65 to 104.4 kilos in 1977/79, an average annual rate of increase of 2.9 kilos. However, there was a tendency for the rate of increase to slow in the 1970s, the annual growth being 2.1 kilos from 1970/72 to 1976/78. We assume that the tendency for the rate of increase to slow will continue, and that on average per capita consumption from 1977/79 to 2010 will rise 1.7 kilos annually. Soybean consumption per capita in 2010 then would be 159 kilos.

Projections of Grains and Soybeans to 2010

The projections of U.S. production, exports and domestic use of wheat, feedgrains and soybeans are shown in appendix table 2-7.

Projections of Cotton

These projections are based on a USDA study by Collins, Evans, and Barry (1979), with certain adjustments by us. Their projections are for U.S. production, export, and domestic use of cotton in 1985 and 1990. The basic assumptions underlying the projections are as follows (Collins and coworkers, 1979, p. 4):

(1) There will be no major wars in the world between now and 1990.

(2) Cotton is viewed as a homogeneous commodity; no distinction is made among various staples and grades.

(3) A synthetic noncellulosic fiber combining the best properties of both cotton and current state of the art synthesis will not become widely available.

(4) A low-cost process that permits cotton to assume all the easy-care and abrasion-resistant properties of synthesis will not become widely available.

(5) There will be no technological advances in cotton production that permit dramatic yield increases worldwide (such as multi-adversity seed variety).

(6) Any sustained worldwide food production shortages that might occur would not result in a major shift of land from cotton production to food production.

Appendix Table 2-7. U.S. Production, Domestic Use, and Exports of Wheat, Feedgrains and Soybeans

(million metric tons)

Crop	1979/80[a]	2010
Wheat		
Production	60.8	98
Domestic use	22.1	28
Exports	38.4	70
Feedgrains		
Production	207.7	354
Domestic use	153.0	187
Exports	72.6	167
Soybeans		
Production	56.3	120
Domestic use	24.3	44
Exports	29.3	76

Sources: U.S. Department of Agriculture (USDA), Foreign Agriculture Circular Grains, August 13, 1980, for wheat and feedgrains; and USDA Agricultural Outlook (Washington, D.C., April 1980) for soybeans. In the projections, changes in stocks are assumed to be zero. Export projections were derived by multiplying projections of world trade by U.S. shares. For wheat and feedgrains the projections of world trade are the sums of imports from appendix tables 2-2 and 2-3 (wheat) and 2-2 and 2-5 (coarse grains). World trade in soybeans is from appendix table 2-6. Projections of domestic use are described in the previous section.

[a]Averages for the two years, except for soybeans the years are 1978/ 79. The differences between production and the sum of utilization and exports is the change in stocks. Exports of soybean meal and oil are expressed in soybean equivalents.

(7) The price of manmade fiber will increase relative to the price of an equivalent amount of cotton, compared with price averages over 1974-76.

(8) Government policies on fiber consumption, production and trade will remain unchanged from current policies unless specified otherwise in certain markets.

(9) Cotton and manmade fiber will compete in a mature market between now and 1990, especially in developed regions. This means the relative price of cotton and manmade fiber will become more important in determining the market share of each fiber.

Within this framework of assumptions, two sets of projections to 1990 are developed, one assuming slightly slower growth of world income than in the 1960s and 1970s and the second assuming that income grows 25 percent less than in the 1960s and 1970s. The growth of world demand and supply of fiber is projected in accord with these projections of income growth. Cotton's share of the world fiber market is assumed to continue to decline, as it has for several decades, but at a slower rate than in the 1960s and early 1970s. The main reason for cotton's declining share, according to Collins and coauthors, has been consumer preference for easy-care and abrasion-resistant fabrics competitive with cotton in price.

Projections of world trade in raw cotton are based on analysis of demand-supply balances among developed, developing and centrally planned economies. Under the higher income growth alternative world trade in raw cotton is 20.9 million bales in 1985 and 23.3 million bales in 1990. With the lower income growth alternative trade is 18.8 million bales in both 1985 and 1990. In 1974-76 world trade averaged 18 million bales.

The U.S. share of world trade with the higher income alternative is 27 percent. With the lower income alternative it is 19 percent. In 1974-76 the U.S. share averaged 22 percent. The decline in the U.S. share with slower growth is attributed by Collins and coauthors to the fact that the United States is one of the most price elastic members among major cotton exporting countries for both supply and demand.

Domestic demand for cotton in the United States is projected in two steps: (1) a projection of demand for fiber, based primarily on projections of income and the income elasticity of demand for fiber; (2) a projection of cotton's share of total demand for fiber. The share is expected to decline gradually from 27 percent in 1977 to 22 percent in 1990.

Within the two income growth scenarios, alternative projections of U.S. production, exports and domestic production are made in accordance with different assumptions about ratios of cotton prices to prices of polyester fibers. These price assumptions have little effect on the projections, however (Collins and coauthors, 1979, p. 23).

The various assumptions made by these authors result in a variety of projections of U.S. production, exports and domestic consumption of raw cotton. The production projections for 1990 range from 10.0 million bales

Appendix Table 2-8. Production, Net Exports, and Mill Consumption of Raw
Cotton in the United States

	1976/79 Millions		1990 Millions		2010 Millions	
	480-1b. bales	Metric tons	480-1b. bales	Metric tons	480-1b. bales	Metric tons
Production	12.6	2.7	15.6	3.4	16-18	3.5-3.9
U.S. mill consumption	6.4	1.4	6.8	1.5	np	np
Exports	5.7	1.2	8.8	1.9	np	np

Sources: 1976-79 from U.S. Department of Agriculture, Foreign Agriculture Circular Cotton, FC 17-79 (Washington, D.C., USDA, November 1979).
Figures for 1990 from K. J. Collins, R. B. Evans, and R. D. Barry, World
Cotton Production and Use: Projections for 1985 and 1990, Foreign Agricultural Economic Report no. 154 (Washington, D.C., USDA, 1979). The projections of production and net exports in 1990 were increased by 2.5
million bales, as described in the text. For projections for 2010, see
text.

to 13.4 million bales. We have elected to base our projection on a figure
of 13.1 million for 1990 plus an increment of 2.5 million bales. The 13.1
million bale figure is the 1990 projection of Collins and coauthors (1979)
corresponding to the higher income growth scenario and a cotton-to-
polyester price ratio of 1.15:1. We selected this scenario because it is
in line with the World Bank's projections of world income growth (World
Bank, 1980) which we used in our projections of trade in grains and oilmeal. We added the increment of 2.5 million bales because in discussion
with one of the authors in the summer of 1980, Collins and coauthors (1979)
said they then believed that the 13.1 million bale projections was low by
about that amount. The reason was that the 13.1 million bale figure did
not reflect changed expectations about the growth of demand in the PRC and
Western Europe for U.S. cotton exports. Accordingly, we project U.S. cotton production in 1990 at 15.6 million bales, and obtain our projection
for 2010 by extrapolation. These and related figures are shown in appendix
table 2-8. The projections to 2010 assume that the maximum growth of cotton production after 1990 will be half the annual rate projected by
Collins and coauthors for 1976-79 to 1990. At a minimum cotton production
would grow only .4 million bales from 1990 to 2010.

References

Collins, K. J., R. B. Evans, and R. D. Barry. 1979. World Cotton Production and Use: Projections for 1985 and 1990, Foreign Agricultural Economic Report No. 154 (Washington, D.C., USDA).

Food and Agriculture Organization. 1979. Agriculture: Toward 2000 C 79/24 (Rome, FAO).

Schnittker Associates. 1979. Trade Issues Relating to World Hunger. A report prepared for the Presidential Commission on World Hunger (Washington, D.C., March 30).

Tsui, Amy Ong, and D. J. Bogue. 1978. "Declining World Fertility: Trends, Causes, Implications," Population Bulletin vol. 3, no. 4 (October).

USDA. 1974. The World Food Situation and Prospects to 1985. Foreign Agricultural Economics Report no. 98 (Washington, D.C., December).

_____. 1976. Foreign Agriculture Circular Grains, FG 9-76 (Washington, D.C., May).

_____. 1976. Foreign Agriculture Circular Grains, FG-17-76 (Washington, D.C., August 3).

_____. 1978. Agricultural Statistics (Washington, D.C.).

_____. 1979. Foreign Agriculture Circular FOP 2079 (Washington, D.C., January).

_____. 1979. Foreign Agriculture Circular Cotton, FC 17-19 (Washington, D.C., November).

_____. 1979. Agricultural Outlook (Washington, D.C., December).

_____. 1980. Agricultural Outlook (Washington, D. C., April).

_____. 1980. Foreign Agriculture Circular Grains. FG-23-80. (Washington, D.C., August 13).

_____. 1981. Foreign Agriculture Circular Grains. FG-46-81 (Washington, D.C., December 15).

World Bank. 1978. 1978 World Bank Atlas (Washington, D.C.).

_____. 1980. World Development Report (Washington, D.C.).

Appendixes to Chapter 4

Appendix A.   ANALYSIS OF TRENDS IN CROP YIELDS

Pure Yield and Shift Effects

If $\dfrac{Q^1}{L^1} \div \dfrac{Q^0}{L^0}$ is the weighted average increase of yields for a given crop in period 1 relative to period 0 across a number of states with Q = the quantity of crop output and L = acres of land, then

$$\frac{\dfrac{q_a^1}{\ell a^1} \cdot \dfrac{\ell a^0}{L^0} + \dfrac{q_b^1}{\ell b^1} \cdot \dfrac{\ell b^0}{L^0} + \cdots + \dfrac{q_n^1}{\ell n^1} \cdot \dfrac{\ell n^0}{L^0}}{\dfrac{q_a^0}{\ell a^0} \cdot \dfrac{\ell a^0}{L^0} + \dfrac{q_b^0}{\ell b^0} \cdot \dfrac{\ell b^0}{L^0} + \cdots + \dfrac{q_n^0}{\ell n^0} \cdot \dfrac{\ell n^0}{L^0}}$$

is the pure yield increase, i.e., that part of the total increase in average yields from period 0 to period 1 which is attributable to yield increases in each state. It is found by weighting yields in each state in each period by that state's share of land in the crop in period 0.

$$Q = q_a + q_b + \cdots q_n$$
$$L = \ell a + \ell b + \cdots \ell n, \text{ where}$$

the subscripts refer to states and superscripts to periods.

The pure shift effect is measured in a similar way:

$$\frac{\dfrac{q_a^0}{\ell a^0} \cdot \dfrac{\ell a^1}{L^1} + \dfrac{q_b^0}{\ell b^0} \cdot \dfrac{\ell b^1}{L^1} + \cdots + \dfrac{q_n^0}{Ln^0} \cdot \dfrac{\ell n^1}{L^1}}{\dfrac{Q^0}{L^0}}$$

In this case the only factor affecting yields is change in each state's
share of land in the crop from period 0 to period 1.

We have calculated these measures with yields in period 0 equal to
100. Consequently, the yield effect minus 100 plus the shift effect minus
100 plus an interaction effect (found as a residual) minus 100 equals the
observed percent change in yields.

We calculated these effects for corn, wheat, sorghum, and soybeans
in those states which accounted for 80 to 90 percent of total production
in the period since 1946-1950. Appendix table 4-1A gives the results.
The table shows that, except for soybeans, pure yield effects accounted
for 90 percent or more of observed changes in yields. The shift effect
for yields of soybeans was negative because of a marked relative shift of
land in that crop to Arkansas, Tennessee, Mississippi, and Louisiana,
where yields are less than in the Corn Belt, the other principal producing
region. Even for soybeans, however, the pure yield effect was by far the
most important factor in observed yield changes over most of the period
considered.

Perhaps the most significant aspect of Appendix table 4-1A is the
decline in the rate of increase of yields of corn, sorghum and soybeans,
and the decline in wheat yields, from 1971-1973 to 1977-1979 compared with
the earlier periods shown. This apparent break in trend is generally consistent with the behavior of the index of all crop yields (see table 4-1).
We noted that the behavior of that index was generally consistent with
the slower shift to land-saving technologies which occurred after 1972.
This must also partially explain the behavior of yields of grains and
soybeans. We know, however, that weather has a powerful effect on yields
in any given year, and that it may also affect the trend of yields. Before we can make statements about the effect of land quality and technology on yields, therefore, we must take out the effects of weather.

Pure Yield Factors: Weather

We have estimates of the effects of weather on yields of corn and
soybeans in the Corn Belt and of wheat in the Great Plains. The estimates are based on the work of Louis Thompson of Iowa State University

Appendix Table 4-1A. Pure Yield, Pure Shift, and Interaction Effects on
Observed Yields of Corn, Wheat, Sorghum, and Soybeans

Crop	1946/50 to 1966/70	1966/70 to 1971/73	1971/73 to 1975/76	1971/73 to 1977/79
Corn				
% change in yields	100.7	21.6	-8.2	8.9
Pure yield	91.0	21.5	-9.1	n.a.
Pure shift	10.5	2.1	.8	n.a.
Interaction	-.8	-2.0	.1	n.a.
Wheat				
% change in yields	66.5	15.0	-5.9	-1.8
Pure yield	65.3	15.7	-6.3	n.a.
Pure shift	1.0	-.6	.5	n.a.
Interaction	.2	-.1	-.1	n.a.
Sorghum				
% change in yields	180.0	15.0	-15.9	.9
Pure yield	172.2	14.5	-16.3	n.a.
Pure shift	-1.3	-.1	.1	n.a.
Interaction	9.1	.6	.3	n.a.
Soybeans				
% change in yields	31.0	6.0	-1.4	8.1
Pure yield	39.2	5.7	-.6	n.a.
Pure shift	-7.8	.6	-.6	n.a.
Interaction	-.4	-.3	-.2	n.a.

Note: n.a. means not available because time did not permit calcula-
tion of these numbers. However, visual inspection of the data indicates
that the pure yield effect accounted for most of the change in yields for
each crop between 1971/73 and 1977/79.

Sources: Data on yields are from U.S. Department of Agriculture,
Agricultural Statistics, various years. Procedures for calculating pure
yield, pure shift, and interaction effects are described in the text.

(National Oceanic and Atmospheric Administration, 1973; and Thompson, 1977). For corn and soybeans the estimates refer to Ohio, Indiana, Illinois, Iowa, and Missouri. For wheat, the estimates are for North and South Dakota, Nebraska, Kansas, and Oklahoma.

For corn and soybeans Thompson's analysis covers the years 1891-1976. For wheat the period is 1893-1973. For corn the share of the five states in total land in corn nationwide was about 40 percent in the late forties and fifties, rising to a little over 50 percent in the early 1970s. The share of these states in total soybean land was about 75 percent in the late forties, falling to about 50 percent in the early 1970s. The share of the five wheat states in total wheat land was about 50 percent in the forties and fifties, rising to about 65 percent in the early 1970s. Thompson presents graphs showing that yields of each crop in each set of states were more variable than national average yields but that the trends of yields in the states were very similar to national trends.

Thompson relies on a regression of yields against time and weather. Time is measured in years and weather in deviations of monthly precipitation and temperature values in particular crop season months from long-term average values of precipitation and temperature for those months. Time is assumed to measure technology. The analysis gives annual estimates of yields for each state, reflecting technology and weather. The estimates for each state are weighted by the harvested area of the state in a given year to get estimates for groups of states. The weighting procedure eliminates the effects on yields of shifts in the share of each state in total harvested area. The effects of shifts within states to land of differing quality would show up, however, as an effect of technology.

To get at the effects of weather over time, Thompson holds technology constant by selecting a given year, 1973, and then substituting weather variables in his equations for each year in his time series. The result is an estimate for each year of what yields would have been, given 1973 "technology" and the weather for that year.

We calculated an index of these technology-adjusted yields, setting yields with "normal" (i.e., long-term average) weather equal to 100. Annual differences in this yield index reflect, in principle, the effects

of year-to-year differences in the weather. By dividing this yield index
for each year into actual yields, we get an estimate of yields for each
year which is independent of the effect of weather. Movements in the
yield series derived in this way thus reflect only the effect of technology.

Appendix tables 4-2A, 4-3A, and 4-4A show for corn, soybeans and
wheat, the index of weather effects on yields, actual yields, actual
yields adjusted for weather, and two sets of weather-adjusted trend values
for yields. Examination of weather-adjusted yields in the tables indicates clearly that those series include influences other than "technology."
With respect to corn, for example, we know that in 1970 yields were greatly
reduced by a kind of blight, an occurrence independent of weather, or in
any case, not picked up by Thompson's method of measuring weather effects.
Nevertheless, for all three crops, weather-adjusted yields are less variable from year to year than actual yields, the expected result. In any
case, we have nothing better. Accordingly, we assume that the weather-
adjusted series roughly approximate the effects of technology on the trend
of yields of corn, wheat, and soybeans.

We ask the following question about the data in the tables: is the
behavior of actual yields of corn, soybeans and wheat from 1972 to 1980
consistent with the hypothesis that, after adjustment for weather, the
trend of yields in that period was the same as the pre-1972 trend? If the
answer is yes, we would conclude that there was no change in the effect of
land quality or technology on the growth of yields after 1972. If the
answer is no, we would conclude that yield growth had been affected by
changes in land quality, or in technology, or both.

We are handicapped by the lack of data which would permit us to adjust yields for weather for years after 1972, as the tables indicate. For
wheat we can make the adjustment only for 1973 and for corn and soybeans
only for 1973, 1974, 1975 and 1976. For these years the differences between weather-adjusted yields and trend values of weather-adjusted yields
do not suggest a break in trend after 1972. That is, for each crop the
differences observed in 1973-1976 are consistent with those observed in
1950-72.

Appendix Table 4-2A. Corn: Index of Effects of Weather on Yields, Actual
Yields, Weather-Adjusted Yields, and Trend Values of Weather-Adjusted
Yields

Year	Index of weather effects[a]	Yields (bu per acre)		Trend of weather-adjusted	
		Actual	Weather-adjusted	Arithmetic	Logarithmic
1950	96.8	49.0	50.6	47.1	50.0
51	95.2	48.1	50.5	49.4	51.1
52	92.8	55.7	60.0	51.7	53.3
53	95.6	51.0	53.4	54.0	55.0
54	92.5	51.1	55.3	56.3	56.7
1955	90.7	51.8	57.1	58.7	58.6
56	97.9	58.4	59.7	61.0	60.4
57	97.4	58.7	60.3	63.3	62.4
58	99.5	64.7	65.0	65.6	64.4
59	99.4	64.0	64.4	67.9	66.4
1960	97.1	64.2	66.1	70.3	68.6
61	102.9	74.2	72.1	72.6	70.8
62	103.6	77.0	74.4	74.9	73.0
63	102.2	80.8	79.0	77.2	75.4
64	98.1	72.6	74.0	79.5	77.8
1965	101.5	85.9	81.1	81.9	80.3
66	96.6	81.3	84.2	84.2	82.8
67	97.8	87.8	89.8	86.5	85.5
68	99.1	88.6	89.4	88.8	88.2
69	100.9	94.0	92.1	91.1	91.0
1970	98.1	77.5	79.0	93.5	94.0
71	97.1	100.6	103.6	95.8	97.0
72	101.4	107.7	106.2	98.1	100.1
73	99.3	101.1	101.8	100.4	103.3
74	78.8	76.7	97.3	102.7	106.6
1975	91.5	97.3	106.3	105.1	110.0
76	93.0	98.0	105.4	107.4	113.5
77		95.6		109.7	117.1
78		109.9		112.0	120.9
79		122.0		114.3	124.7
1980		103.6		116.7	128.7

Note: Yields are given for Ohio, Indiana, Illinois, Iowa, Missouri.

Sources: The index of weather effects 1950-1973 is derived from National Oceanic and Atmospheric Administration, The Influence of Weather and Climate bn United States Grain Yields: Bumper Crops or Droughts, A Report to the Administrator of NOAA (Washington, D.C., December 14, 1973) p. 26. The indexes for 1974, 1975, and 1976 are from Louis M. Thompson, "Climatic Variability and World Grain Production," paper presented at a meeting of the American Seed Trade, Kansas City, Mo., November 8, 1977. Actual Yields: 1950-1978 from USDA, Agricultural Statistics, various years, 1979-1980 from USDA, 1980, Crop Production CrPr 2-2 (9-80) (Washington, D.C. Crop Reporting Board, September 11). Weather-adjusted yields are actual yields divided by the index.

[a]"Normal" yields = 100, 1973 "technology." Values above 100 indicate better than "normal" weather.

Appendix Table 4-3A. Soybeans: Index of Effects of Weather on Yields,
   Actual Yields, Weather-Adjusted Yields, and Trend Values of Weather-
   Adjusted Yields

Year	Index of weather effects[a]	Yields (bu per acre)			
		Actual	Weather-adjusted	Trend of weather-adjusted Arithmetic	Logarithmic
1950	99.8	23.1	23.1	21.8	22.1
51	96.6	23.1	23.9	22.2	22.4
52	97.7	23.1	23.6	22.7	22.8
53	91.9	19.6	21.3	23.1	23.2
54	97.1	22.1	22.7	23.5	23.5
1955	93.6	21.1	22.6	23.9	23.9
56	96.8	23.1	23.8	24.3	24.3
57	98.2	24.7	25.1	24.8	24.7
58	105.7	26.8	25.3	25.2	25.1
59	97.8	25.8	26.3	25.6	25.5
1960	97.5	25.2	25.9	26.0	25.9
61.	101.6	27.6	27.2	26.4	26.3
62	102.6	26.8	26.1	26.9	26.7
63	99.8	27.8	27.9	27.3	27.1
64	96.8	24.7	25.5	27.7	27.5
1965	101.0	27.3	27.0	28.1	28.0
66	98.0	27.3	27.9	28.5	28.4
67	97.0	26.5	27.3	29.0	28.9
68	99.3	31.0	31.2	29.4	29.3
69	102.7	31.5	30.7	29.8	29.8
1970	98.4	30.2	30.7	30.2	30.2
71	97.2	31.6	32.5	30.6	30.7
72	100.6	32.3	32.1	31.1	31.2
73	102.6	30.7	29.9	31.5	31.7
74	84.0	24.9	29.6	31.9	32.2
1975	97.6	33.1	33.9	32.3	32.7
76	93.7	29.9	31.9	32.7	33.2
77		35.9		33.2	33.6
78		34.0		33.6	34.3
79		36.4		34.0	34.8
1980		32.7		34.4	35.4

Note: Yields are given for Ohio, Indiana, Illinois, Iowa, Missouri.

Sources: The index of weather effects 1950-1973 is derived from National Oceanic and Atmospheric Administration, The Influence of Weather
and Climate on United States Grain Yields: Bumper Crops or Droughts,
Report to Administrator of NOAA (Washington, D.C., December 14) p. 28.
The indexes for 1974, 1975, and 1976 are from Louis M. Thompson, "Climatic Variability and World Grain Production," paper presented at a
meeting of the American Seed Trade, Kansas City, Mo. Actual Yields:
1950-78 from USDA Agricultural Statistics, various years; 1979-80 from
USDA, 1980, Crop Production CrPr 2-2(9-80)(Washington, D.C., Crop
Reporting Board, September 11). Weather-adjusted yields are actual
yields divided by the index.

[a]"Normal" yields = 100, 1973 "technology." Values above 100 indicate better than "normal" weather.

Appendix Table 4-4A. Wheat: Index of Effects of Weather on Yields,
   Actual Yields, Weather-Adjusted Yields and Trend Values of Weather-
   Adjusted Yields

| Year | Index of weather effects[a] | Yields (bu per acre) | | Trend of weather-adjusted | |
		Actual	Weather-adjusted	Arithmetic	Logarithmic
1950	96.3	14.0	14.5	14.1	14.7
51	93.7	13.4	14.3	14.8	15.2
52	98.6	16.7	17.0	15.5	15.7
53	92.4	12.6	13.6	16.3	16.3
54	97.9	14.3	14.6	17.0	16.8
1955	95.7	15.3	15.9	17.7	17.4
56	87.5	16.1	18.4	18.4	18.0
57	89.9	19.1	21.2	19.1	18.6
58	110.1	26.8	24.3	19.8	19.3
59	91.9	18.1	19.7	20.6	19.9
1960	104.8	24.9	23.7	21.3	20.6
61	94.7	21.6	22.8	22.0	21.3
62	98.9	23.1	23.3	22.7	22.1
63	93.2	21.0	22.5	23.4	22.8
64	95.3	22.9	24.0	24.1	23.6
1965	96.8	24.1	24.9	24.9	24.4
66	94.1	22.3	23.7	25.6	25.2
67	90.5	21.5	23.7	26.3	26.1
68	100.7	26.5	26.3	27.0	27.0
69	108.7	29.7	27.3	27.7	27.9
1970	95.7	28.9	30.2	28.4	28.9
71	104.1	31.9	30.7	29.2	29.9
72	101.6	30.5	30.0	29.9	30.9
73	108.5	32.1	29.6	30.6	32.0
74		23.9		31.3	33.1
1975		26.7		32.0	34.2
76		25.9		32.7	35.4
77		27.4		33.4	36.6
78		28.7		34.2	37.9
79		32.6		34.9	39.2
1980		27.6		35.6	40.5

   Note: Yields are given for North Dakota, South Dakota, Nebraska,
Kansas, Oklahoma.

   Sources: The index of weather effects is derived from National
Oceanic and Atmospheric Administration, The Influence of Weather and Climate on United States Grain Yields: Bumper Crops or Droughts, Report to
the Administrator of NOAA (Washington, D.C., December 14, 1973). Actual
yields: 1950-1978 from USDA Agricultural Statistics, various years; 1979-
1980 from USDA, 1980, Crop Production CrPr 2-2(9-80)(Washington, D.C.,
Crop Reporting Board, September 11). Weather-adjusted yields are actual
yields divided by the index.

[a]"Normal" yields = 100, 1972 "technology." Values above 100 indicate better than "normal" weather.

Because we cannot adjust for weather after 1976 (1973 in the case of wheat), we adopt a different approach in assessing yield behavior in the more recent years. We ask the following question: given the historical effects of weather on yields, what is the probability that weather could have caused the observed differences between actual yields and the trend values of weather-adjusted yields?[1] In asking this question we assume that weather was the main cause of annual differences between actual yields and the trend values of weather-adjusted yields in 1950-1972, and that other causes were randomly distributed over the period. This of course is at best an approximation to the truth. We believe it is a sufficiently good approximation, however, to provide insights into the behavior of actual yields in recent years. If, for example, in every recent year actual yields differed from trend values of weather-adjusted yields by more than the amount observed in any year in the historical record, we would suspect strongly that something had happened to the _trend_ of weather-adjusted yields in recent years.

For those years for which we are unable to adjust for weather, we take the difference between actual yields and the trend value of weather-adjusted yields and apply a chi-square test to judge whether the observed differences could be due to weather, given the effects of weather on yields in 1950-1972. For wheat, the years considered are 1974-1980. For corn and soybeans they are 1977-1980. For wheat the chi-square test is unequivocal--the observed yield differences could not be due to weather alone.[2] Not only were the differences greater than would be expected on the basis of weather experienced from 1950-1972, but observed yields were less than the trend value of weather-adjusted yields in each year 1974-1980. It is highly probable that there was a tilt downward in the effect of land quality or technology on wheat yields after 1973. We return to this below.

---

[1]In considering the question, we use differences between actual yields and the arithmetic trend values of weather-adjusted yields. For all three crops the arithmetic trend gave as good a fit to the weather-adjusted yield data as the logarithmic trend.

[2]The test indicated that the probability of differences as large as those observed if weather were responsible was much less than 1 in 100.

For corn the chi-square test gives a probability of about 6 percent that the differences in 1977-1980 between actual yields of corn and the trend values of weather-adjusted yields were attributable to weather alone. Although small, this probability is high enough to indicate we cannot rule weather out as the main reason for the yield differences.

The chi-square test indicates that the differences between actual soybean yields and trend values of weather-adjusted yields in 1977-1980 could have been caused by the weather.[3] There is no convincing evidence, therefore, of a downward tilt in the effect of either land or technology on soybean yields. Indeed if there was a change in quality of soybean land or technology it more likely gave an upward tilt, since actual yields exceeded trend yields in three of the four years 1977-1980. However, we assume no change in land quality or technology effects on the trend of soybean yields in the Corn Belt. Instead we concentrate on corn and wheat, attempting to isolate the effects of land quality and technology on yields of these crops after 1972.

Pure Yield Factors: Land Quality

It is reasonable to assume that for any given level of production farmers will use that land which gives the highest net return per acre. This is not necessarily the land with the highest physical yield per acre, but we expect there is a high correlation between net return and physical yield. Without putting too fine a point on it, we assume accordingly that if land is taken out of production, average yields will rise. If more land is brought into production, average yields will fall.

Precise measurement of the effect on yields of changes in the amount of land is hampered--in fact made impossible--by the absence of data showing yield differences between land in production at any given time and increments or decrements of land between that time and any other. In the absence of such data, we have improvised to try to establish limits to the effects of additional land on observed yields of corn and wheat. For each crop we ask, what would production have been in 1977/80 on land in produc-

---

[3]The probability that weather alone was responsible is about 15 in 100.

tion in 1972 if the pre-1972 trend of yields on that land had continued? This hypothetical production on this land in 1977/80 is subtracted from actual production in those years, and the difference is divided by the increment of land between 1972 and 1977/80. This gives an estimate of yields on the increment assuming that yields on land in production in 1972 increased between that year and 1977/80 at the pre-1972 rate. We then ask whether the yield estimate on the increment of land seems reasonable, for example, by comparison with estimated yields on land in production in 1972. If yield on the increment seems too low, the implication is that the trend of yields on land in production in 1972 was less than the pre-1972 trend.

This procedure obviously is very rough. Aside from the arbitrary nature of the assumptions about yields on the increment of land, it also is assumed that the gross and net changes in land after 1972 are the same. For example, if land in corn is 50 million acres in 1972 and 60 million acres in 1977/80, the procedure assumes that the same 50 million acres is in production in both periods with an addition of 10 million acres after 1972. This may not be the case. It is possible, for example, that 5 million of the 50 acres is shifted to some other use while 15 million acres formerly in some other use are shifted to corn. We really should measure the effects on yields of both the 15 million acres and the 5 million acres, but the procedure we have used cannot do it. The problem probably is not too serious, however, at least for the period considered. Harvested land in wheat and corn in the states we are analyzing increased by a net of about 13 million acres from 1972 to 1977/80. The amount of land previously in these crops which was shifted to other uses in this period probably was small compared to the additional 13 million acres. Hence the difference between the net and the gross increment likely was small.

Changes in the amount of land in wheat and corn in the states of interest here may have significantly affected the pre- and post-1972 trends of yields of those crops. Wheat in particular may have been affected. Between 1946-1950 and 1972 the amount of land in wheat in the five wheat states--North Dakota, South Dakota, Nebraska, Kansas and Oklahoma--declined from 36.8 million acres to 25.2 million acres. If, in 1946-1950, yields on the 11.6 million acres subsequently taken out of production

were 90 percent of yields on the land remaining in production, then removal of the lower yielding land would have contributed a little over 3
percent to the growth of weather-adjusted wheat yields from 1946-1950 to
1972.[4] Between 1972 and 1977-1980, an average of 7.9 million additional
acres of land in these states were brought into wheat production, an
increase of almost one-third. This reversal in the declining trend of
land in wheat may have significantly reduced wheat yields in 1977-1980
below what they otherwise would have been.

There was no significant change between 1946-1950 and 1972 in the
amount of land in corn in the five Corn Belt states--Ohio, Indiana, Illinois, Iowa and Missouri. Consequently, under the assumptions of our procedure, changes in the amount of land could not have affected the trend
of corn yields in that period. Between 1972 and 1977-1980, however, the
average amount of land in corn in these states increased by 5.4 million
acres, or about 18 percent. This may have depressed somewhat the growth
of corn yields between 1972 and 1977-1980.

We have used the procedure described above to estimate the effect on
wheat and corn yields in 1977-1980 of bringing more land under these crops
after 1972. The results are in appendix table 4-5A. The purpose of the
exercise is to estimate the contribution of additional land to the differences between actual yields and weather-adjusted trend yields in 1977-1980
(i.e., the differences between lines 1 and 2 in the table. If all of the
differences were attributable to the additional land, then yields on that
land would have been as shown in line 7, i.e. yields on the increment
of wheat land would have averaged 11.9 bushels per acre in 1977-80,
and yields on the increment of corn land would have averaged 76.6
bushels. Had the pre-1972 trend of weather adjusted yields continued,
wheat yields in 1977/80 would have been 34.5 bushels and corn yields would
have been 113.2 bushels (line 1 in appendix table 4-5A). Thus, estimated
yields on the increment of wheat land were 34 percent of trend yields and
on corn land they were 68 percent of trend. Weisberger (1969) estimated
that in the 1960s yields on set-aside wheat land were about 90 percent of

---

[4]P. Weisgerber (1969) estimated that yields on wheat land held in reserve in the late 1960s were about 90 percent of land in wheat production.

Appendix Table 4-5A. Analysis of the Effects of Changing the Amount of
Land on Yields of Wheat and Corn

	Wheat 1977/80	Corn 1977/80
1) Yields if pre-1972 trend in weather-adjusted yields had continued	34.5	113.2
2) Actual yields (bu.)	29.1	107.7
3) Production in 1977/80 on 1972 land if yields had grown at pre-1972 rate (mill. bu.)	869.4	3452.6
4) Actual production in the indicated years (mill. bu.)	963.2	3866.4
5) Line 4 minus line 3 (mill. bu.)	93.8	413.8
6) Increment of land from 1972 to the indicated years (mill. acres)	7.9	5.4
7) Yield in 1977/80 on increment of land (line 5 divided by line 6 -bu.)	11.9	76.6

yields on land in production. For corn the corresponding percentage was 82. By this standard the estimated 1977/80 yield on additional wheat land clearly seems too low. That is, it is unlikely that yield on the additional land would have been only 11.9 bushels in those years if yields on land in production in 1972 had been 34.5 bushels, the amount given by pre-1972 trend. The implication is that the post-1972 yield trend on that land was less than the pre-1972 trend.

The analysis of the effects of weather and additional land on wheat yields indicates that neither can explain the difference between actual yields and the trend values of weather adjusted yields in 1977/80. But what about the combined effect of these two factors? We cannot estimate this precisely because we lack data about the yield effect of additional land. However, if we assume that yields on the additional land were 90 percent of yields on 1972 land (basing this on Weisberger, 1969) a test is possible. We ask the question, could the differences in 1977/80 be-

tween trend values of weather adjusted wheat yields and actual yields
adjusted for additional land be explained by weather? Applying the chi-
square test as before, the answer is no.[5] That is, the combined effect
of weather and additional land does not explain the failure of wheat
yields after 1972 to grow as rapidly as indicated by the pre-1972 trend.

Application of a similar test to corn indicates that the difference
between actual corn yields in 1977/80 and the trend value of weather
adjusted yields could be attributed to the combined effect on yields of
weather and additional land. In other words, after allowing for the combined effect of these two factors, it appears that there was no significant difference between actual corn yields and the trend values of weather
adjusted yields in 1977/80.

The analysis of corn yields has focused on the Corn Belt because in
that region we were able to adjust yield data for effects of weather.
Corn yields in the rest of the nation, however, moved much as they did in
the Corn Belt after 1972, that is the rate of increase slowed considerably. The amount of harvested land in corn outside the Corn Belt increased 8.3 million acres, or 30 percent, from 1972 to 1977-1980, substantially more than in the Corn Belt both in absolute amount and in percentage. Since our analysis of Corn Belt yields suggests that increased acreage was responsible for a major part of the slowdown in yield growth after
1972, it seems likely that much of the slower growth of yields outside the
Corn Belt also may be attributable to additional acreage. However, as
pointed out below, per acre use of fertilizer on corn increased more slowly
after 1972, and this too may have depressed the growth of corn yields.

Pure Yield Factors: Technology

The analysis to this point indicates that in the Corn Belt weather
alone could account for differences in 1977/80 between actual yields and
weather adjusted trend yields of soybeans, and for corn the combined effect of weather and additional land would suffice. For these two crops
in that major producing area there does not appear to have been a post-

---

[5]The probability that the differences could be explained by weather
is less than 1 percent.

1972 break in the effect of technology on yields. Weather and additional
land do not explain the slow growth of wheat yields in the plains states
after 1972, however. A possibility is that technological factors may
have been responsible. This is true also for the growth of corn yields
outside the Corn Belt.

Fertilizer

Appendix table 4-6A shows fertilizer use data for corn and wheat.
The data are consistent with substantially slower growth in yields of both
crops after 1970-1972. From 1964-1966 to 1970-1972, wheat yields increased 1.05 bushels per acre per year. From 1970-1972 to 1977-1979,
wheat yields actually declined slightly. Corn yields grew 3.2 bushels
per acre per year from 1964-1966 to 1971-1973 and by 1.4 bushels annually
from 1971-1973 to 1977-1979.[6]

In the discussion of the effects of weather and additional land on
corn yields in the Corn Belt, we concluded that those two factors together
could have been responsible for the slower growth of corn yields after
1972. The data in appendix table 4-6A suggest, however, that the slower
growth in amount of fertilizer applied per acre in corn may also have contributed, if not in the Corn Belt then in other corn growing areas.

Appendix table 4-6A suggests that fertilizer played a role in the
slower growth of wheat yields after 1972, but the case is not strong.
Over one-third of wheat land received no fertilizer at all in 1977/79,
and the percentage in earlier years was less. Technological factors
other than fertilizer could have been quite significant in the growth of
wheat yields.

Irrigation[7]

Irrigated production of corn and wheat grew rapidly from 1950 to
1977--4748 percent for corn and 470 percent for wheat--reflecting both an

---

[6]We used 1971-1973 for corn instead of 1970-1972 because 1970 yields
were severely depressed by an attack of leaf blight.

[7]This discussion is based on Frederick (1982). We deal only
with western irrigation, since to date the contribution of eastern irrigation to the growth of yields of crops of interest here has been negligible.

Appendix Table 4-6A. Fertilizer Use on Corn and Wheat in the United
States

	Acres receiving %			Amount per re-ceiving acre (1b)			Amount per harvested acre (1b)			
	N	P	K	N	P	K	N	P	K	Total
Corn										
1964-1966	88	82	76	73	22	39	64	18	30	112
1970-1972	95	89	84	111	66	68	105	59	57	221
1977-1979	96	88	82	130	68	82	125	60	67	252
Wheat										
1964-1966	48	37	15	30	13	25	14	5	4	23
1970-1972	60	43	16	42	34	37	25	15	6	46
1977-1979	63	42	18	53	37	39	33	16	7	56

Note: N = nitrogen; P = phosphorous; K = potassium

Source: U.S. Department of Agriculture, Fertilizer Situation, FS-2,
FS-5, and FS-10 (Washington, D.C., January 1972, December 1974, and
December 1979). The data begin with 1964.

increase in irrigated acreage in these crops and rising yields. However,
total irrigated production of corn and wheat was small relative to dryland production. Thus irrigation accounted for only 28 percent of the
national increase in corn production and for 12 percent of the increase in
wheat production. Yields of irrigated corn and wheat exceed dryland
yields, and irrigated yields increased from 1950 to 1977. Consequently,
irrigation contributed to rising yields of corn and wheat--both because
of the expansion of irrigated acreage relative to dryland acreage in these
crops and because of rising yields on the irrigated land. Nonetheless,
because irrigated acreage in and production of corn and wheat was small
relative to dryland acreage and production, the contribution of irrigation to the increase in national yields between 1950 and 1977 was small--
13 percent for corn and 7 percent for wheat.

Reliable data on changes in the amount of irrigated land in corn and
wheat since 1972 are not available. However, most of the expansion of
irrigated land in these crops for at least ten or fifteen years before

1972 was in the Plains region, particularly Nebraska and Kansas. Discussion with people in the region who follow irrigation suggests little slowdown in the rate of expansion since the early 1970s. If this is true, irrigation would not likely have contributed to slower growth of national corn and wheat yields after 1972. The effect on yields of even relatively large changes in irrigated land would be difficult to detect in any case because of the relatively small contribution of irrigation to the change in national yields.

Summary

There is no evidence of a change in the trend of yields of soybeans in the five Corn Belt states after 1972. The observed differences between actual yields and trend values of weather-adjusted yields could easily have resulted from variations in the weather.

In the Corn Belt actual corn yields fell short of the trend values of weather-adjusted yields in three of the four years 1977-1980. Unfavorable weather and the bringing in of less productive land explain most if not all of the yield shortfall. National average corn yields also grew more slowly after 1972 and slower growth in fertilizer use per acre of corn land probably contributed to this as well as weather and less productive land. There is no evidence that irrigation contributed to slower growth of corn yields.

Actual wheat yields in the Great Plains were less than the trend values of weather-adjusted yields in every year from 1974 to 1980. Weather could not have caused this nor could the combined effects of weather and less productive land. Per acre applications of fertilizer to wheat land increased more slowly after 1972 than before, but the percentage of wheat land receiving fertilizer is too small for this factor to have been of major importance. There is no evidence to suggest that slower growth in irrigated wheat acreage contributed to the yield shortfall. We are left without an adequate explanation of the failure of wheat yields to grow after 1972 in accordance with the pre-1972 trend of weather-adjusted yields.

Appendix B. TRENDS IN TOTAL AGRICULTURAL PRODUCTIVITY

Appendix table 4-1B shows rates of growth in total agricultural productivity over various periods between 1950 and 1972 and between 1972 and
1980. As the years advanced from 1950 to 1960, the annual rate of increase in productivity declined. From 1960 to 1972 it was stable to
slightly increasing, and from 1972 to 1980 it again declined. It can be
argued that between 1950 and 1972 American farmers were engaged in exploit-
ing the high-productivity potential of a new technology consisting basic-
ally of fertilizers, pesticides, high-yielding seed varieties, mechanical
power substituting for labor and animals, and to a lesser extent, irrigation. If this statement captures the essence of what was happening, then
the decline and subsequent leveling off from 1950 to 1972 in the rate of
increase of productivity is plausible. The argument runs as follows.
When the new technology began to spread after World War II there was a
large difference in productivity between it and the technologies then in
use. Consequently the rapid spread of the new technology produced a fast
increase in productivity. By about 1960, however, the substitution of the
new for the old technology was substantially completed. Subsequent increases in productivity, therefore, reflected incremental improvements in
the new technology rather than wholesale substitution of the new for the
old. In this circumstance a slower rate of increase in productivity after
about 1960 would be expected.[1]

---

[1] By 1960 high-yielding varieties of corn and sorghum had almost 100
percent replaced lower yielding varieties previously in use, and a major
part of the substitution of mechanical power for animal power had been
completed by the mid-1950s. (Inputs of mechanical power and machinery
increased at an annual rate of 6 percent from 1946 to 1955, but by only
.24 percent annually from 1955 to 1972. Fertilizer consumption, however,
grew even faster after 1960 than it did before that date.)

Appendix Table 4-1B. Average Annual Rates of Growth of Total Agricultural Productivity in the United States

	Percentage	Index Points
1950-1972	2.05	1.80
1955-1972	1.91	1.77
1960-1972	1.62	1.59
1965-1972	1.64	1.69
1972-1980	.97	1.08

Note: The growth rates are derived from least squares analysis of
annual index numbers of total productivity. Logarithms of these numbers
were used to calculate percentage rates of change.

Source: U.S. Department of Agriculture, Economic Indicators of the
Farm Sector: Production and Efficiency Statistics, ERS Statistical Bulletin no. 679 (Washington, D.C., January 1982).

Whether this account explains the apparent decline in productivity
growth after 1972 is not clear, and the issues cannot be pursued here.
Indeed the discussion trends in agricultural productivity is vexed by
serious problems in the measurement of total productivity. There is
strong evidence that failure to include quality improvements in inputs
resulted in significant underestimates of increases in the quantities of
inputs, thus overestimating the increase in productivity.[2] Perhaps more
important, the vast technological changes in American agriculture since
the end of World War II make it difficult to construct an index which
unambiguously measures the growth of inputs; consequently measurement of
productivity change also is ambiguous. The input index currently used by
the USDA weights inputs by their average prices in 1971-1973. Wages were
much higher in those years relative to prices of other inputs than at the
end of World War II, but the amount of labor was sharply reduced relative
to those inputs, particularly fertilizer and other agricultural chemicals.
Use of 1971-1973 weights gives relatively great importance to labor and
hence dampens the rise of the total input index. If price weights of

[2]National Academy of Sciences (1975), pp. 28-29. Also, USDA (Feb.
1980), especially pp. 28-32.

Appendix Table 4-2B. Average Annual Percentage Rates of Increase in Production and Selected Inputs in U.S. Agriculture

	1960-1972	1972-1980
Production		
Crops	1.6	2.7
Animal Products	1.7	.7
Total	1.6	2.1
Inputs		
Fertilizer	7.3	3.7[a]
Labor	-4.5	-3.3[a]
Machinery	.3	3.5[a]
Farm real estate	.2	.0[a]
Harvested cropland	-.2	1.8

Note: Rates are derived from least squares analysis of logarithms of indexes of inputs and production.

Source: U.S. Department of Agriculture, Economic Indicators of the Farm Sector: Production and Efficiency Statistics, ERS Statistical Bulletin no. 679 (Washington, D.C., January 1982).

[a]1972-1979.

earlier years were used—say 1946-1949—labor would weigh less and other inputs more, and the input index would rise much more steeply than the one used by the USDA. The index of total productivity, of course, would also rise less.[3]

There is no generally satisfactory solution to this index number problem, and the USDA is not to be faulted for not having solved it. The point is that the ambiguity in the productivity index requires that movements in it be interpreted cautiously, and that conclusions about those movements be checked for consistency with other indicators of agricultural performance.

A question arises also about the behavior of the index specifically in the period after 1972. Two features of appendix table 4-2B prompt the

---

[3]A study by Barton and Durost, cited in National Academy of Sciences (1975, p. 31), indicated that for the period 1940-1942 to 1956-1958, the index of total inputs increased 23 percent when prices of 1935-1939 were used as weights. When 1957 prices were used the index declined 1 percent.

question: (1) the fact that most of the growth in total output after 1972 was attributable to crops, and (2) the contrasting behavior in this period of farm real estate and harvested cropland. Farm real estate includes all land in farms, service buildings, grazing fees, and repairs on service buildings. It thus includes harvested cropland. The amount of harvested cropland increased after 1972 even though farm real estate was unchanged because some farmland previously held idle or in other uses was shifted to crops. It is farm real estate which is included in the index of inputs, but it was the increase in harvested cropland which accounted for much of the increase in crop production, and hence total production, after 1972. It appears, therefore, that because of the way land is handled in the index of inputs, the rise in the index after 1972 was underestimated, with consequent overestimation of the rise in productivity.

References

Economic Report to the President. 1981. Government Printing Office, Washington, D.C.

Frederick, Kenneth. 1982. "Irrigation and the Adequacy of Agricultural Land," in P. Crosson (ed.), The Cropland Crisis: Myth or Reality, Johns Hopkins Press for Resources for the Future (Baltimore).

National Academy of Sciences. 1975. Agricultural Production Efficiency (Washington, D.C.).

National Oceanic and Atmospheric Administration. 1973. The Influence of Weather and Climate on United States Grain Yields: Bumper Crops or Droughts, a Report to the Administrator of NOAA, December 14 (Washington, D.C.).

Thompson, Louis M. 1977. "Climatic Variability and World Grain Production," paper given at a meeting of the American Seed Trade, Kansas City, Mo., November 8. Thompson is Associate Dean of Agriculture, Iowa State University.

USDA. Various Years. Agricultural Statistics. U.S. Government Printing Office (Washington, D.C.).

_____. 1972. Fertilizer Situation. FS-2, January (Washington, D.C.).

_____. 1974. Fertilizer Situation. FS-5, December (Washington, D.C.).

_____. 1979. Fertilizer Situation. FS-10, December (Washington, D.C.).

_____. 1980. Basic Statistics: 1977 National Resources Inventory, Revised (Washington, D.C., Soil Conservation Service, Feb.).

_____. 1980. Measurement of U.S. Agricultural Productivity: A Review of Current Statistics and Proposals for Change, Technical Bulletin no. 1615 (Washington, D.C., Economics, Statistics and Cooperatives Service, February).

_____. 1982. Economic Indicators of the Farm Sector: Production and Efficiency Statistics, ERS Statistical Bulletin no. 679 (Washington, D.C., January).

Weisberger, P. 1969. Productivity of Converted Cropland. USDA, Economic Research Service, ERS-398 (Washington, D.C.).

_____. 1980. Crop Production. CrPr 2-9 (9-80) (Washington, D.C., Crop Reporting Board, September 11).

Appendix to Chapter 5

MATERIAL RELATING TO THE DEMAND FOR
AND SUPPLY OF CROPLAND

Regional Shares of Production

As indicated in chapter 5, the regional shares of production of main
crops on the whole have been quite stable, and that weighed heavily in
our projections of shares. There are, however, some currents of change
which we have taken into account, and these are described here.

We noted in chapter 3 that there is considerable potential for increased production of irrigated soybeans in the Mississippi Delta, and
some potential for expansion or irrigated corn and soybean production in
the Southeast. Accordingly, we have projected an increase in the share
of the Delta's production of soybeans and in the Southeast's share of
both soybeans and corn. Most of the increase would occur after 1985.
The Southeast is allocated the greatest percentage increase in production
of soybeans because that region has more land in pasture and forest with
potential for conversion to cropland than the Delta, both in absolute
amount and in relation to its present cropland base.[1] The increased
shares of the Delta and the Southeast in soybean production would come
at the expense of the Corn Belt, although that region would continue to
be by far the nation's leading producer of soybeans. However, as will be
pointed out below, our projections of crop production and yields in the
Corn Belt imply heavy pressure on the region's supply of cropland by 2010,
including land now in forest, pasture, and range with potential for conversion to cropland. Indeed, our projections of crop production and
yields imply that the amount of land in crops in the Corn Belt would
exceed the present and potential supply by 2010 unless the region's

_____

[1]Data on potential cropland are from USDA (February 1980). These
data are presented in table 5-5.

shares of production of some crops decline. We project a small decline
in the share of feedgrain production, but we expect a much larger decline
in the share of soybeans because of the demonstrated potential for expanded
soybean production in both Delta and the Southeast.

The study by Shulstad and coauthors (1980) of irrigation potential in
the Mississippi Delta indicates that double-cropping of wheat and soybeans
would be one of the more profitable options. If soybean production in the
Delta expands as we expect, it is likely that wheat production would increase also. This is reflected in table 5-1 in the sharp relative increase in wheat production projected for the Delta.

We expect continued growth in the share of the Southern Plains in
wheat production, reflecting a process evident for some years, and consistent with the region's relative abundance of land in range and pasture
with potential for conversion to cropland. The principal losers in shares
of wheat production would be the Corn Belt and Northern Plains, although
the later region would remain by far the most important wheat producer in
the nation.

We expect the Southern Plains to increase its share of cotton production also, continuing the trend evident over the last couple of decades.
However, future conditions are likely to favor the Southern Plains (mainly
Texas) relative to Arizona and New Mexico as well as relative to the Southeast and the Mississippi Delta. Most cotton production in the two western
states is irrigated and will be exposed, therefore, to the increasing constraints on western irrigation noted in chapter 3. By contrast, only 40
to 45 percent of the cotton land in Texas is irrigated. Moreover, research in Texas has developed cotton varieties especially well adapted to
growing conditions in that state, and which reduce the pest management
problem relative to its dimension in the Southeast and Delta.[2]

Apart from factors already discussed, our projections of regional
shares of production also were influenced by comparison of two runs of
Iowa State University's model of the agricultural economy of the United

---

[2]There is more on this in the discussion in chapter 6 of trends in
insecticide use on cotton.

States. The runs were done especially for this study.[3] In one of the runs we supplied our projections of crop production and a preliminary allocation of production among the ten USDA producing regions. In the second run we used the same nationwide projections of production, but asked the model to allocate the production among regions. The results of the two runs are shown in appendix table 5-1. If we accept the model's allocation as that which would occur as farmers seek to minimize costs of production and transport among markets, then we would conclude that the regional distribution of production resulting from run 2 is more likely than that resulting from our initial distribution imposed upon the model, represented by run 1. We do not accept the model's distribution as the ultimate truth, but we do believe it should be given some weight. Accordingly, where, in our judgment, our initial distribution differed significantly from the model's distribution we adjusted the initial distribution to bring it closer to the model's. These adjustments were for feedgrains among the Great Lakes, Ohio, Upper Mississippi, and Missouri regions; for wheat among the Upper and Lower Mississippi and the Arkansas-White-Red; and for oilmeal (soybeans) among the Lower Mississippi, the Missouri, and the Arkansas-White-Red. We judged the distribution of cotton in the two runs to be similar enough to need no adjustment.

The various adjustments are incorporated in the regional distribution shown in appendix table 5-1.

Yields

We expect prices and productivities of inputs to favor more land-using technologies, as they have since the early 1970s. In general, therefore, we take yield behavior in the 1970s as a guide in projecting future yields. For feedgrains, however, we assumed that yields would not grow as rapidly as experience in the 1970s might suggest. The main reason is that

---

[3]The model was developed by Earl Heady and associates. It is a linear programming type which allocates production of major crops among regions in such fashion as to minimize production and transportation costs. The model's structure does not permit allocation of production among the 10 USDA production regions we have used. Instead the regions are those shown in appendix table 5-1.

Appendix Table 5-1. Percentage Distribution of Production of Selected
Crops Among River Basins in the United States in 2010

Region or Basin	Run 1 Feed-grains	Run 1 Wheat	Run 1 Oil-meal[b]	Run 1 Cotton	Run 2 Feed-grains	Run 2 Wheat	Run 2 Oil-meal[b]	Run 2 Cotton
New England	a	0	0	0	0	0	0	0
Middle Atlantic	2.6	.7	3.1	0	1.4	2.2	3.2	0
South Atlantic-Gulf	6.7	.5	15.8	8.0	4.1	1.9	13.0	9.2
Great Lakes	8.8	5.2	5.7	0	16.1	2.7	1.8	0
Ohio	14.1	4.2	16.7	0	10.7	4.3	20.0	0
Tennessee	.4	a	3.7	0	.3	0	1.5	0
Upper Mississippi	27.1	9.4	24.7	0	37.1	2.4	22.0	0
Lower Mississippi	.1	1.3	14.4	1.2	1.5	7.7	8.6	1.4
Souris-Red-Rainey	.1	8.6	2.9	0	0	8.5	2.0	0
Missouri	25.6	31.3	4.9	0	16.3	34.5	14.2	0
Arkansas-White-Red	7.3	20.4	3.2	9.6	5.5	14.4	9.2	11.1
Texas-Gulf	3.4	1.0	3.9	48.7	4.0	1.7	3.6	44.6
Rio Grande	.5	.2	.2	6.5	.8	a	.1	5.3
Upper Colorado	.1	1.2	0	0	.1	1.5	0	0
Lower Colorado	.1	.6	.1	5.3	a	0	.1	4.5
Great Basin	.1	0	0	0	0	.5	0	0
Columbia-North Pacific	.6	13.2	0	0	.3	14.4	0	0
California-South Pacific	2.1	2.2	.6	20.7	2.2	3.2	.6	23.9

Note: In run 1 the distribution among the USDA producing regions was
imposed by RFF. The model then allocated among the regions in the table.
In run 2 the model allocated national production among the regions.

Source: Iowa State University model of the U.S. agricultural economy.

[a] Less than .1 percent.

[b] Virtually all of this is soybean.

we expect the price of fertilizer in the future to behave differently relative to the price of corn than it did in 1973/1979. From 1973 to 1975, the price of nitrogen rose sharply relative to the price of corn, and the amount of nitrogen applied per acre of corn declined.[4] From 1975 to 1979, the price of nitrogen decline relative to the price of corn, and per acre applications rose. This pattern must have tended to depress corn yields in 1973/75 and then raise them to 1979. As indicated in chapter 3, we expect the price of nitrogen to rise relative to the price of corn over the next several decades, in contrast to its behavior in 1975/1979, and that the amount of nitrogen applied per acre of corn will rise less rapidly than from 1975 to 1979.[5] Accordingly, we expect corn yields to rise more slowly from 1979 to 2010 than from 1973 to 1979. Since corn currently accounts for about 70 percent of the land in feedgrains, and for an even larger percentage in our projections, the slower growth of corn yields implies slower growth of feedgrain yields.

We expect soybean yields in the Corn Belt, Delta and Southeast to grow more rapidly than national yields. Our analysis in chapter 4 of soybean yields in the Corn Belt indicated that the growth of yields in that region did not slow after 1972, as they did nationwide, and we see no reason to expect different performance in the future. Consequently we projected soybean yields in the Corn Belt to grow to 2010 in accord with the arithmetic trend of weather-adjusted yields as shown in appendix table 4-3A.

Our discussion of eastern irrigation indicated considerable expansion potential for irrigated soybeans in the Mississippi Delta and, to a lesser extent, in the Southeast. To reflect this we projected a significant increase in soybean yields in those regions, using yield estimates in Shulstad and coauthors (1980) as a guide. Their study also indicates greatly increased potential for double-cropping wheat with soybeans in the Delta, and this looks promising also in the Southeast. We increased national wheat yields over what the trend since 1973 would suggest to reflect this potential.

---

[4]See table 3-4 for the ratio of nitrogen prices to corn prices.

[5]Projections of fertilizer are developed in chapter 6.

Demand for Cropland

Derivation of the demand for land in main crops is described in chapter 5. Here we discuss the projections of demand for land in other crops and other uses of cropland.

Since the late 1960s, the amount of land in "other crops" has been quite stable at around 90 to 95 million acres. About two-thirds of this land consistently is in hay. Land in hay reached a peak of 77.6 million acres in 1944, then declined steadily to 60.9 million acres in 1968. It has varied narrowly around 60 million acres ever since. We have assumed that land in hay will be 60 million acres in 1985, but that it will decline to 50 million acres in 2010 as hay has increasing difficulty competing with other higher-valued uses of the land. We assume that non-hay "other crops" will occupy the same amount of land in 1985 and 2010 as in 1977. To do better than this would require more analysis of future demand for these crops, and of their yields, than we are in a position to make.

In each year from 1958 to 1977, crops failed on from 5 million to 10 million acres, the average being 7.4 million acres. In projecting 10 million failed acres, the figure for 1977, we may be somewhat on the high side, but any reasonable alternative would make no significant difference in our projections of total demand for cropland.

In the same 20-year period just mentioned, land in fallow varied between 30 million acres and 41 million acres, the average being 33.9 million acres. Virtually all of the land in fallow is in the main wheat growing areas. Fallowing restores soil moisture after harvest, and is essential to economical production of wheat in the semi-arid West. The record suggests that this imperative sets a minimum of about 30 million acres for land in fallow. We have used that figure in both 1985 and 2010.

Data for idle cropland are available only for agricultural census years. That for the four most recent years are shown in appendix table 5-2. The high figures for 1964 and 1969 reflect the policy then in effect of encouraging farmers to take land out of production; the low figure for 1974 reflects the abandonment of that policy and the influence of high-crop prices in stimulating intensive use of the land. It is significant that despite this stimulus there still were some 20 million acres of idle

Appendix Table 5-2. Idle Cropland in the United States for Selected Years

Year	Millions of acres
1954	33
1964	52
1969	51
1974	20

Source: U.S. Department of Agriculture, Agricultural Statistics (Washington, D.C., GPO, 1973 and 1977). 1978 agricultural census data for idle cropland were not available at this writing.

cropland in the country in 1974. According to land use specialists at the USDA, some cropland inevitably is idle every year for reasons that vary widely from place to place, and 20 million acres probably is close to the minimum amount of such land. That is the basis for our projection of idle land.

References

Shulstad, R.M., R.D. May, B.E. Herrington, and J.M. Erstine. 1980. Expansion Potential for Irrigation Within the Mississippi Delta Region (Fayetteville, Department of Agricultural Economics and Rural Sociology, University of Arkansas).

USDA. Various Years. Agricultural Statistics (Washington, D.C., GPO).

_____. 1980. Basic Statistics: 1977 National Resources Inventory (Washington, D.C., Soil Conservation Service, February).

Index

For Product Safety Concerns and Information please contact our EU
representative GPSR@taylorandfrancis.com
Taylor & Francis Verlag GmbH, Kaufingerstraße 24, 80331 München, Germany

www.ingramcontent.com/pod-product-compliance
Ingram Content Group UK Ltd.
Pitfield, Milton Keynes, MK11 3LW, UK
UKHW022255051225
465792UK00006B/48